Mathematical Modelling

Kluwer Texts in the Mathematical Sciences

VOLUME 28

A Graduate-Level Book Series

The titles published in this series are listed at the end of this volume.

Mathematical Modelling

Case Studies and Projects

by

Jim Caldwell
City University of Hong Kong,
Kowloon, Hong Kong, P.R. China

and

Douglas K.S. Ng
City University of Hong Kong,
Kowloon, Hong Kong, P.R. China

KLUWER ACADEMIC PUBLISHERS
DORDRECHT / BOSTON / LONDON

A C.I.P. Catalogue record for this book is available from the Library of Congress.

ISBN 978-90-481-6566-7   e-ISBN 978-1-4020-1993-7

Published by Kluwer Academic Publishers,
P.O. Box 17, 3300 AA Dordrecht, The Netherlands.

Sold and distributed in North, Central and South America
by Kluwer Academic Publishers,
101 Philip Drive, Norwell, MA 02061, U.S.A.

In all other countries, sold and distributed
by Kluwer Academic Publishers,
P.O. Box 322, 3300 AH Dordrecht, The Netherlands.

Printed on acid-free paper

Jim Caldwell would like to dedicate this book to:

Ann, Janice, Clare and Simon Caldwell, and

Kei Shing Ng would like to dedicate this book to

his parents for giving him life and support.

TABLE OF CONTENTS

PREFACE

Over the past decade there has been an increasing demand for suitable material in the area of mathematical modelling as applied to science, engineering, business and management. Recent developments in computer technology and related software have provided the necessary tools of increasing power and sophistication which have significant implications for the use and role of mathematical modelling in the above disciplines. In the past, traditional methods have relied heavily on expensive experimentation and the building of scaled models, but now a more flexible and cost effective approach is available through greater use of mathematical modelling and computer simulation. In particular, developments in computer algebra, symbolic manipulation packages and user friendly software packages for large scale problems, all have important implications in both the teaching of mathematical modelling and, more importantly, its use in the solution of real world problems.

Many textbooks have been published which cover the art and techniques of modelling as well as specific mathematical modelling techniques in specialist areas within science and business. In most of these books the mathematical material tends to be rather tailor made to fit in with a one or two semester course for teaching students at the undergraduate or postgraduate level, usually the former. This textbook is quite different in that it is intended to build on and enhance students' modelling skills using a combination of case studies and projects. The material for these case studies and projects is reflected in the leading author's lengthy teaching, research and industrial experience and falls within the following three major areas:

(1) Ordinary Differential Equations (ODEs);

(2) Partial Differential Equations (PDEs);

(3) Optimization.

Also, it could be argued that these three areas are key areas in which mathematical modelling can prove effective. Provided that the model formulation is successful in producing an ODE, PDE or a primal/dual linear programming (L.P.) problem, there is a strong possibility of successful solution because of the wealth of appropriate analytical and numerical techniques. In recent times, model validation becomes more possible through the use of computer algebra systems and specialist software packages.

The book is aimed primarily at the postgraduate level and should prove suitable for students of mathematics or closely related disciplines which require mathematical modelling, e.g., science, engineering, business and management. Also the book should prove attractive to industrialists with modelling interests. Some of the case studies have been formulated from the industrial and research experience of the leading author, whereas the projects are more typical of what can be expected from a final year undergraduate or a Master's student.

In preparing the text, the authors have tried to use their experience of teaching mathematical modelling to both undergraduate and graduate students in a wide range of areas including mathematics and computer science and disciplines in engineering, science, business and management. An important aspect of the book is the use made of scientific computer software packages such as MAPLE for symbolic algebraic manipulations, MATLAB for numerical simulation and the popular linear programming package LINDO.

The book is divided into three main parts with each part containing two major case studies and three projects, followed by suggested problems. Parts I, II and III deal with the three key areas of ODEs, PDEs and Optimization, respectively. This means that the book comprises case studies A1, A2, ..., A6 and projects B1, B2, ... , B9. The case studies are distinguished from the projects in that they tend to be more comprehensive and pitched at a higher level. As such, they should provide excellent teaching material. Having mastered the two case studies and three projects, the students should be in a strong position to tackle the ten suggested problems at the end of each Part of the book. To help the reader, and for consistency, each case study and project is solved using a standard format. Further details are given in the Introduction.

The authors would like to express their thanks to other authors, former and present modelling colleagues and final year undergraduates (too numerous to mention by name) who have helped over the past years to give some of the important modelling ideas contained in this textbook. This includes (in alphabetical order) R. P. Canale, S. C. Chapra, A. Constantinides, L. R. Foulds, R. C. Klekamp, Y. Y. Kwan, M. M. Meerschaert, M. L. Ruwe, R. J. Thierauf and D. Waters.

In the preparation of the actual textbook, the authors would like to thank Marlies Vlot of Kluwer Academic Publishers for all her organizational help. Particular thanks must go to Mr CHONG Wai-san (Mountain) and to Mr CHUI Ka-shing (Daniel) for their help on the computing side and in preparing some of the more difficult computer graphs and figures. Thanks also to Miss CHAN Kit-yee (Jamie) for her word processing help.

Last but not least, a particular vote of thanks must go to Miss LEE Sau-yan (Eva) for skillfully typing part of the manuscript and for work on the optimization case studies and projects.

Dr Jim Caldwell
Mr Kei Shing Ng

INTRODUCTION

The term *model*, as used in this textbook, is understood to refer to the ensemble of equations which describe and interrelate the variables and parameters of a physical system or process. The term *modelling* in turn refers to the derivation of appropriate equations that are solved for a set or system of process variables and parameters. These solutions are often referred to as simulations, i.e., they simulate or reproduce the behaviour of physical systems and processes.

Modelling is practiced frequently in the engineering disciplines and indeed in all physical sciences where it is often known as "Applied Mathematics". However, it has made its appearance in other disciplines as well which do not involve physical processes per se, such as economics, finance and management science.

The textbook is organized to include a number of modelling case studies and projects and it has proved convenient to section the book into the following three parts:

Part I. Ordinary Differential Equations

Part II. Partial Differential Equations

Part III. Optimization

Each Part consists of two major case studies, three worked projects followed by ten suggested problems. In Part I, the ODE case studies and projects are drawn from the areas of mechanical, electrical and civil engineering as well as bioscience. In Part II, the PDE case studies and projects are drawn from the areas of mechanical, chemical/petroleum and civil/environmental engineering. Lastly in Part III, the Optimization case studies and projects involve linear programming and

1

transportation type problems and are drawn from the areas of business, finance and management science.

The detailed layout is as follows:

Part I — Ordinary Differential Equations

Case Study A1. Deterministic Model in Contagious Disease.
Case Study A2. Electromagnetic Forces in High Field Magnet Coils.

Project B1. Mass Balance of a Reactor in Steady State.
Project B2. The Free and Forced Vibration of an Automobile.
Project B3. Cantilever Beam Subjected to an End Load.

Part II — Partial Differential Equations

Case Study A3. Cylindrical and Spherical Solidification in Heat Transfer.
Case Study A4. Elastic Analysis of a Square Plate with Circular Holes.

Project B4. Motion of Fluid Layers.
Project B5. Mass Balance of a Reactor with Time Dependency.
Project B6. Flow through Porous Media.

Part III — Optimization

Case Study A5. Linear Programming Problem Involving Wine Production.
Case Study A6. Transportation Problem Involving Breweries and Hotels.

Project B7. Profit from an Engineering Plant.
Project B8. Optimization of Manufacture of Personal Computers.
Project B9. Air Freight Transportation Problem.

It should be explained that the case studies give a more detailed treatment of the modelling process with the content at a reasonably high level, usually postgraduate. As such, these case studies should provide excellent teaching material for the lecturer. In the projects which follow, the problem to be solved is more clearly defined and, as such, should give students a good understanding, through example, of the techniques which can be used in the modelling process. Having studied the case studies and worked through the projects, readers should be in a strong position to tackle the problems at the end of each Part of the textbook.

It is important to note that, for both efficiency and consistency, a clearly defined format has been used for both the case studies and projects. The intention here is to provide a systematic approach to enable students and industrialists to tackle problems which can be solved using a mathematical modelling approach. The format is as follows:

Title
Summary
1. Background
2. Problem Statement
3. Model Formulation
4. Mathematical/Numerical Solution
5. Model Validation
6. Interpretation and Conclusions
7. Computer Algorithms
8. References and Bibliography
9. Appendices (if any).

Further details of the type of information contained in each section of the case study/project report are given below:

The title and summary should convey to the reader a good idea of the subject matter and what the case study/project involves. Section 1 (Background) contains essential information which is unique to that particular case study/project. Section 2 (Problem Statement) is intended to explain the objectives and to state what is given and what is required. Section 3 (Model Formulation) breaks the problem down and formulates it mathematically in terms of governing equations/inequalities and relevant boundary/initial conditions. Section 4 (Mathematical/Numerical Solution) presents a mathematical solution (if this is possible) and/or, failing that, possible numerical solutions. Section 5 (Model Validation) is key to any model solution and could involve solutions using alternative numerical techniques or the use of the computer, including software packages, to back up the results obtained in the previous section. Section 6 (Interpretation and Conclusions) interprets the results and draws major conclusions giving, in some cases, suggestions for further extensions. Section 7 (Computer Algorithms) discusses and, in some cases, gives details of any important algorithms or software which have been used in the solution techniques. Section 8 (References and Bibliography) lists in alphabetical order important textbooks or, particularly for the case studies, journal references. Finally, Section 9 (Appendices) is only included in cases where there is detailed information which is not considered suitable for the body of the report.

Part I

ORDINARY DIFFERENTIAL

EQUATIONS

Case Study A1

DETERMINISTIC MODEL IN CONTAGIOUS DISEASE

SUMMARY: This case study extends past work by Caldwell and Ram in the study of a deterministic model in the theory of contagious disease. A more realistic model is considered by introducing a third variable, namely, the number of removals in the population. Numerical results are obtained by using the Runge-Kutta-Fehlberg method including error control. Results and graphs are produced to show the effects of variation of the infection rate and the removal rate on the number of removals from the population over long time periods. These results are validated and are shown to agree well with analytical results.

1. Background

Throughout the world numerous people die from infections by serious diseases like black death, smallpox, tuberculosis, etc. Although some of these fatal diseases are gradually disappearing from our lives, many widespread diseases still exist resulting in the death of millions of people. Medical workers and health authorities have devoted substantial efforts and resources into trying to predict and control the spread of diseases of many types. Here mathematicians can be of invaluable assistance and play an important role which will help to decide how resources are allocated. Hence it is important to be able to predict how a disease will develop and spread (if at all). Mathematical investigations in the theory of epidemics are therefore important in predicting the development and spread of diseases.

The first mathematical model which was used to study the effectiveness of inoculation against smallpox was developed by the Swiss mathematician Bernouille in the 17th century. Since the mid-19th century mathematical theories have developed much faster with the increased understanding and control of contagious diseases. At the present time there are in excess of 500 primarily mathematical references to the population theory of infectious disease.

Inevitably different epidemic models will be applied to different cases, e.g., if the population is sufficiently large and fixed, no people are immune, all infected people are infectious and vice versa, then we can use a deterministic model. However, when dealing with small groups the probability element must be taken into account and so stochastic models are used.

2. Problem Statement

In this case study we consider a deterministic model along the lines of that discussed by Bailey [1,2], and extend the work of Caldwell and Ram [3]. Here a homogeneously mixed group of individuals of size $n+a$ was considered under the assumptions that initially a individuals are infected with the remaining n individuals all being susceptible, but not yet infected. This leads to the classical simple model.

Let time t be the independent variable, $I(t)$ and $S(t)$ be continuous, where

$$S(t) = \text{number of susceptibles at time } t$$

$$I(t) = \text{number of infectives at time } t.$$

Under the major assumption that the rate of occurrence of new infections is proportional to both the number of infectives and the number of susceptibles, we can write

$$I(t+\Delta t) = I(t) + \beta I(t)S(t)\Delta t, \tag{A1.1}$$

where $\beta =$ infection rate (or contact rate). In the limit as $\Delta t \to 0$, this yields

$$\frac{dI(t)}{dt} = \beta I(t)S(t) \tag{A1.2}$$

with initial conditions $S(0) = n$, $I(0) = a$.

In addition, since the total population size is always $n+a$, and all individuals are either susceptible or infected, it is clear that $S(t)+I(t) = n+a$ for all t and it follows that

$$\frac{dI(t)}{dt} = \beta I(t)\{n + a - I(t)\} \cdot \tag{A1.3}$$

Caldwell and Ram [3] considered the above model and applied the Runge-Kutta-Fehlberg algorithm to obtain highly accurate results. In this case study we consider a more realistic and general model of an epidemic by introducing a third variable $R(t)$ to represent the number of individuals who are removed from the affected population at a given time t, either by isolation, recovery and consequent immunity, or death. The numerical tool used in the solution is the Runge-Kutta-Fehlberg algorithm discussed in [3]. Numerical results will be obtained by varying the infection rate and removal rate and conclusions drawn. Comparisons are also made with analytical results.

3. Model Formulation

The previous simple model (A1.3) is now extended to make it more realistic by taking into account the removal of infectives from circulation by death or isolation. The following assumptions are made for this model:

- The removals include infectives who are isolated, dead or recovered and immune;
- The immune or recovered removals enter a new class which is not susceptible to disease.

In addition to the variables used in the previous simple model in Section 1, let $R(t) =$ the number of removals at time t and $\gamma =$ the removal rate, such that we have now

$$I(t) + S(t) + R(t) = n,$$

when n is the total size of the community.

The basic differential equations are constructed as follows. Obviously, if the infectives are removed from the community, they will not be in contact with the susceptibles. Therefore, the number of susceptibles is only proportional to both the number of infectives and the number of susceptibles, and so we have

$$\frac{dS(t)}{dt} = -\beta S(t)I(t) \cdot \tag{A1.4}$$

However, the removals should be considered in the differential equation for the number of infectives we discussed previously. This means that equation (A1.2) should be modified to

$$\frac{dI}{dt} = \beta S(t)I(t) - \gamma I(t) \cdot \tag{A1.5}$$

The equation for the number of individuals who are removed from the infectives with removal rate γ then becomes

$$\frac{dR(t)}{dt} = \gamma I(t) \cdot \tag{A1.6}$$

At the start of the epidemic, when $t = 0$, we assume that there are no removals, a very small number of infectives, I_0, and the remaining population is susceptible, S_0, which is approximately equal to n. Thus, at $t = 0$, (S, I, R) take the values $(S_0, I_0, 0)$. For convenience, we make use of $\mu = \gamma/\beta$, as the *relative removal rate*.

Note that from equation (A1.5), it follows that unless $\mu < S_0$ there will not be an epidemic as $[dI/dt]_{t=0}$ is required to be greater than zero. On the other hand, for the case $\mu > S_0$, the number of infectives will be increasing. Therefore, the relative removal rate, $\mu = S_0$, gives a threshold density of susceptibles.

4. Mathematical/Numerical Solution

4.1 MATHEMATICAL SOLUTION

We now consider the analytical solution of equations (A1.4)-(A1.6). By eliminating I from equations (A1.4) and (A1.6), we have

$$\frac{dS}{dR} = -\frac{S}{\mu} \cdot$$

Integration yields

$$\ln S = -\frac{R}{\mu} + c,$$

leading to

$$S = A\exp(-R/\mu) \cdot$$

At $t = 0$,

$$S = S_0 \exp(-R/\mu)$$

and thus,

$$S = S_0 = A.$$

Since $I = n - S - R$, equation (A1.6) becomes

$$\frac{dR}{dt} = \gamma\{n - R - S_0 \exp(-R/\mu)\}.$$

As $\exp(-R/\mu) = 1 - (R/\mu) + (R/\mu)^2/2! - (R/\mu)^3/3! + \cdots$, we expand the right-hand side of the above equation as far as the term in R^2 to give

$$\frac{dR}{dt} = \gamma\{n - R - S_0[1 - (R/\mu) + (R/\mu)^2/2]\}$$

or

$$\frac{dR}{dt} = \gamma\left[n - S_0 + \left(\frac{S_0}{\mu} - 1\right)R - \frac{S_0}{2\mu^2}R^2\right]. \tag{A1.7}$$

The above equation (A1.7) is soluble by standard methods to give the analytical solution:

$$R = \frac{\mu^2}{S_0}\left\{\frac{S_0}{\mu} - 1 + \alpha\tanh\left(\frac{\gamma\alpha t}{2} - \psi\right)\right\}, \tag{A1.8}$$

where

$$\alpha = \left\{\frac{2S_0}{\mu^2}(n - S_0) + \left(\frac{S_0}{\mu} - 1\right)^2\right\}^{1/2} \tag{A1.9}$$

and

$$\psi = \tanh^{-1}\frac{1}{\alpha}\left(\frac{S_0}{\mu} - 1\right). \tag{A1.10}$$

The epidemic curve is therefore

$$\frac{dR}{dt} = \frac{\gamma\alpha^2\mu^2}{2S_0}\operatorname{sech}^2\left(\frac{1}{2}\gamma\alpha t - \psi\right) \tag{A1.11}$$

and the graph is a symmetric bell-shaped curve. This corresponds to the fact that in many actual epidemics the number of new cases reported daily increases to a maximum and then dies away.

4.2 NUMERICAL SOLUTION

Highly accurate results have been obtained by Caldwell and Ram [3] in using the Runge-Kutta-Fehlberg (R-K-F) algorithm for the simple two variable model. Also this method is very efficient in that it only requires six evaluations per step while arbitrary Runge-Kutta methods of order four and five used together require at least 10 functional evaluations per step. This method is now used to solve the three variable model.

Suppose there is an epidemic of total size 100,000 and there are 99,999 susceptibles and therefore only one infective. Also, we assume the infection rate of the disease is 0.000 009 while the removal rate is 0.89. The R-K-F program is run for 2000 days with a tolerance of one as a longer time is needed if we are interested in what the maximum number of removals could be in the epidemic. So, the differential equation (A1.7) can be used with $R(0) = 0$ and $0 \le t \le 2000$, where

n = total size of the community = 100,000,

S_0 = initial number of susceptibles = 99,999,

a = initial number of infectives = 1,

β = infection (contact) rate = 9×10^{-6},

γ = removal rate = 0.89.

For this case the computer program was run using a minimum and maximum step size of 0.05 and 1.0, respectively. Computed results were obtained by using the R-K-F method at time intervals of 10 days and compared with the analytical solution. Results tabulated at time intervals of 100 days from $t = 0$ to $t = 1500$ are shown in Table A1.1. Clearly there is very close agreement between the actual solution R_i and the numerical solution w_i. This is easily checked by examining the values of the absolute error $|R_i - w_i|$ and the percentage relative error, namely $|(R_i - w_i)/R_i| \times 100\%$. The following conclusion can be drawn from the results tabulated at 10 day intervals:

Maximum absolute error = 4.731×10^{-4} (at $t = 320$),

where numerical solution $w_{320} = 1,195.528\,211$

and analytical solution $R_{320} = 1,195.528\,684$.

Maximum percentage relative error = 7.023×10^{-5} (at $t = 10$),

where numerical solution $w_{10} = 9.358\ 457$

and analytical solution $R_{10} = 9.358\ 464$.

Figure A1.1 shows both the numerical and analytical solutions for the number of removals in the population over a 2000 day period. We note that, initially, the total number of removals grows exponentially and then the growth rate decreases and levels out eventually. Also the errors are larger when the graph shown in Figure A1.1 is of steepest descent.

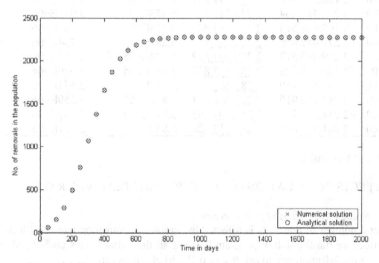

Figure A1.1. Plot of the number of removals from the population against time in days.

From the results we find that the total number of removals becomes almost fixed at $R \approx 2,281$ when $t \approx 1,140$ days. This means that the total number of removals becomes steady after this time and no more individuals would be removed. These values can be verified from the analytical solution (A1.8) by taking the limit as $t \to \infty$. This gives

$$R_\infty = \frac{\mu^2}{S_0}\left(\frac{S_0}{\mu} - 1 + \alpha\right). \tag{A1.12}$$

Clearly the computed solution $R_\infty = 2,281.319\ 143$ is very close to the numerical solution $w_{2000} = 2,281.319\ 057$.

Table A1.1. Comparison of analytical results with numerical results using R-K-F method for the case $\beta = 9 \times 10^{-6}$.

t_i	Analytical solution R_i	Numerical solution w_i	Absolute error $\mid R_i - w_i \mid$	Percentage Error $\mid (R_i - w_i)/R_i \mid \times 100\%$
0	0.000 000	0.000 000	0.000E+00	0.000E+00
100	149.501 525	149.501 424	1.015E−04	6.794E−05
200	493.512 304	493.512 009	2.951E−04	5.979E−05
300	1 068.276 717	1.068.276 252	4.649E−04	4.352E−05
400	1 657.318 561	1.657.318 151	4.098E−04	2.472E−05
500	2 024.085 368	2.024.085 121	2.468E−04	1.219E−05
600	2 186.950 018	2.186.949 881	1.376E−04	6.294E−06
700	2 248.311 964	2.248.311 876	8.853E−05	3.937E−06
800	2 269.973 516	2.269.973 447	6.935E−05	3.055E−06
900	2 277.442 896	2.277.442 833	6.227E−05	2.734E−06
1000	2 279.997 577	2.279.997 517	5.973E−05	2.620E−06
1100	2 280.868 890	2.280.868 831	5.882E−05	2.579E−06
1200	2 281.165 780	2.281.165 722	5.850E−05	2.564E−06
1300	2 281.266 910	2.281.266 851	5.838E−05	2.559E−06
1400	2 281.301 354	2.281.301 295	5.834E−05	2.557E−06
1500	2 281.313 084	2.281.313 026	5.833E−05	2.557E−06

5. Model Validation

5.1 EFFECTS OF VARIATION OF INFECTION AND REMOVAL RATES

5.1.1 Results with different infection rates
The total number of removals is affected by varying the infection rate. To illustrate this effect we run the R-K-F program keeping all the initial values unchanged and changing the infection rate to (a) $\beta = 1 \times 10^{-5}$, (b) $\beta = 0.95 \times 10^{-5}$.

Case (a) $\beta = 1 \times 10^{-5}$
Results are presented in Table A1.2 at 10 day time intervals from $t = 0$ to 200 days. Again there is good agreement between the analytical and numerical results. The following conclusion can be drawn from the results tabulated at 10 day intervals over the time period $t = 0$ to 500 days.

Maximum absolute error = 1.879×10^{-2} (at $t = 70$),

where numerical solution $w_{70} = 9,365.666\ 247$

and analytical solution $R_{70} = 9,365.685\ 043$.

Table A1.2. Comparison of results for the case $\beta = 10^{-5}$, $\gamma = 0.89$.

t_i	Analytical solution R_i	Numerical solution w_i	Absolute error $\|R_i - w_i\|$	Percentage Error $\|(R_i - w_i)/R_i\| \times 100\%$
0	0.000 000	0.000 000	0.000E+00	0.000E+00
10	16.209 911	16.209 840	7.148E−05	4.409E−04
20	64.785 375	64.785 098	2.771E−04	4.277E−04
30	209.390 560	209.389 697	8.627E−04	4.120E−04
40	631.537 848	631.535 390	2.458E−03	3.892E−04
50	1 796.787 864	1 796.781 479	6.385E−03	3.553E−04
60	4 571.924 954	4 571.911 454	1.349E−02	2.952E−04
70	9 365.685 043	9 365.666 247	1.879E−02	2.006E−04
80	14 371.606 824	14 371.592 115	1.470E−02	1.023E−04
90	17 477.397 105	17 477.389 976	7.128E−03	4.078E−05
100	18 830.717 545	18 830.714 888	2.657E−03	1.411E−05
110	19 328.485 926	19 328.485 025	9.012E−04	4.663E−06
120	19 499.925 792	19 499.925 483	3.094E−04	1.586E−06
130	19 557.622 727	19 557.622 607	1.208E−04	6.179E−07
140	19 576.888 550	19 576.888 488	6.280E−05	3.208E−07
150	19 583.304 812	19 583.304 767	4.511E−05	2.303E−07
160	19 585.439 805	19 585.439 766	3.980E−05	2.032E−07
170	19 586.150 011	19 586.149 973	3.823E−05	1.952E−07
180	19 586.386 239	19 586.386 201	3.777E−05	1.928E−07
190	19 586.464 810	19 586.464 772	3.763E−05	1.921E−07
200	19 586.490 943	19 586.490 905	3.760E−05	1.919E−07

Maximum percentage relative error = 4.410×10^{-4} (at $t = 10$),

where numerical solution $w_{10} = 16,209\ 840$

and analytical solution $R_{10} = 16,209\ 911$.

Also as $t \to \infty$ we can use equation (A1.12) to find the total number of removals: $R_\infty = 19,586.503\ 929$ which compares well with the numerical values $w_{200} = 19,586.490\ 905$ and $w_{500} = 19,586.503\ 929$.

Case (b) $\beta = 0.95 \times 10^{-5}$

Results have been obtained at 10 days intervals over the time period $t = 0$ to 1000 days and those results up to $t \to 200$ days are presented in Table A1.3. Again the agreement is good and the following conclusion can be drawn.

Table A1.3. Comparison of results for the case $\beta = 9.5 \times 10^{-6}$, $\gamma = 0.89$.

t_i	Analytical solution R_i	Numerical solution w_i	Absolute error $\mid R_i - w_i \mid$	Percentage Error $\mid (R_i - w_i)/R_i \mid \times 100\%$
0	0.000 000	0.000 000	0.000E+00	0.000E+00
10	12.191 596	12.191 642	4.564E−05	3.743E−04
20	34.372 877	34.373 006	1.291E−04	3.758E−04
30	74.633 184	74.633 464	2.805E−04	3.759E−04
40	147.392 692	147.393 247	5.542E−04	3.760E−04
50	277.863 919	277.864 955	1.036E−03	3.729E−04
60	508.583 073	508.584 939	1.866E−03	3.669E−04
70	906.690 215	906.693 436	3.221E−03	3.553E−04
80	1 565.402 148	1 565.407 392	5.243E−03	3.349E−04
90	2 583.158 735	2 583.166 567	7.832E−03	3.032E−04
100	4 000.419 593	4 000.429 893	1.029E−02	2.574E−04
110	5 712.802 687	5 712.814 211	1.152E−02	2.017E−04
120	7 459.792 424	7 459.803 240	1.081E−02	1.449E−04
130	8 960.010 323	8 960.018 891	8.568E−03	9.563E−05
140	10 069.216 213	10 069.222 178	5.964E−03	5.923E−05
150	10 801.858 404	10 801.862 199	3.794E−03	3.513E−05
160	11 250.407 054	11 250.409 340	2.286E−03	2.032E−05
170	11 512.376 969	11 512.378 311	1.342E−03	1.165E−05
180	11 661.182 029	11 661.182 811	7.827E−04	6.712E−06
190	11 744.374 422	11 744.374 884	4.613E−04	3.927E−06
200	11 790.472 016	11 790.472 297	2.808E−04	2.381E−06

Maximum absolute error = 1.152×10^{-2} (at $t = 110$),

where numerical solution $w_{110} = 5,712.814\ 211$

and analytical solution $R_{110} = 5,712.802\ 687$.

Maximum percentage relative error = 3.760 109 (at $t = 40$),

where numerical solution $w_{40} = 147.393\ 247$

and analytical solution $R_{40} = 147.392\ 692$.

Also as $t \to \infty$ we can again use equation (A1.12) to find $R_\infty = 11,846.856\ 367$ which compares well with the numerical value $w_{1000} = 11,846.856\ 422$.

To assess the effect of variation of infection rate on the total number of removals the results for the cases $\beta = 1 \times 10^{-5}$, $\beta = 0.95 \times 10^{-5}$ and $\beta = 0.9 \times 10^{-5}$ have been presented in Figure A1.2 over a time period of 2000 days. Clearly the gradients of the three curves are different, with the higher the infection rate the steeper the curve. Also the higher the infection rate the shorter the time period required for the number of removals to level out. Clearly larger infection rates lead to higher values of R_∞.

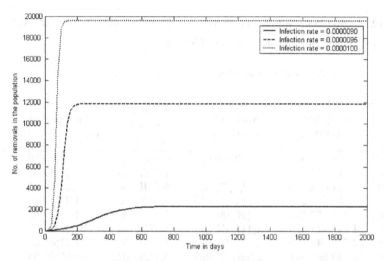

Figure A1.2. Plot of the number of removals from the population against time in days for different infection rates.

5.1.2   Results with different removal rates
Now we consider the effect of varying the removal rate. As before, we keep all the initial values in Section 4.2 unchanged.

Case (a) $\gamma = 0.8$
Results are presented in Table A1.4 at 10 day time intervals from $t = 0$ to 200 days. The agreement between the analytical and numerical results is good. The following conclusion can be drawn from the results tabulated at 10 day intervals over the time period $t = 0$ to 500 days.

Maximum absolute error = 6.438×10^{-2} (at $t = 80$),

where numerical solution $w_{80} = 10,829.004\ 276$

and analytical solution $R_{80} = 10,828.939\ 887$.

Maximum percentage relative error $= 1.273 \times 10^{-3}$ (at $t = 30$),

where numerical solution $w_{30} = 151.747\,165$

and analytical solution $R_{30} = 151.745\,232$.

Also as $t \to \infty$ we can use equation (A1.12) to find the total number of removals:

$R_\infty = 19,759.500\,365$ which compares well with the numerical values
$w_{200} = 19,759.401\,240$ and $w_{500} = 19,759.500\,468$,

Table A1.4. Comparison of results for the case $\beta = 9 \times 10^{-6}$, $\gamma = 0.8$.

t_i	Analytical solution R_i	Numerical solution w_i	Absolute error $\lvert R_i - w_i \rvert$	Percentage Error $\lvert (R_i - w_i)/R_i \rvert \times 100\%$
0	0.000 000	0.000 000	0.000E+00	0.000E+00
10	13.742 277	13.742 450	1.738E−04	1.264E−03
20	51.028 010	51.028 658	6.484E−04	1.270E−03
30	151.745 232	151.747 165	1.933E−03	1.273E−03
40	420.582 778	420.588 098	5.320E−03	1.264E−03
50	1 115.924 436	1 115.938 149	1.371E−02	1.228E−03
60	2 776.948 840	2 776.980 087	3.124E−02	1.125E−03
70	6 089.866 641	6 089.921 967	5.532E−02	9.084E−04
80	10 828.939 887	10 829.004 276	6.438E−02	5.946E−04
90	15 163.389 840	15 163.436 375	4.653E−02	3.068E−04
100	17 778.647 495	17 778.671 171	2.367E−02	1.331E−04
110	18 982.042 132	18 982.052 203	1.007E−02	5.305E−05
120	19 466.405 400	19 466.409 395	3.995E−03	2.052E−05
130	19 650.734 475	19 650.736 053	1.577E−03	8.029E−06
140	19 719.376 802	19 719.377 455	6.533E−04	3.313E−06
150	19 744.731 393	19 744.731 699	3.064E−04	1.552E−06
160	19 754.068 507	19 754.068 684	1.768E−04	8.953E−07
170	19 757.503 188	19 757.503 317	1.284E−04	6.503E−07
180	19 758.766 128	19 758.766 239	1.104E−04	5.591E−07
190	19 759.230 444	19 759.230 548	1.037E−04	5.252E−07
200	19 759.401 139	19 759.401 240	1.012E−04	5.126E−07

Case (b) $\gamma = 0.85$

Results have been obtained at 10 days intervals over the time period $t = 0$ to 800 days and those results up to $t = 200$ days are presented in Table A1.5. Again the agreement is good and the following conclusion can be drawn.

Maximum absolute error $= 7.635 \times 10^{-4}$ (at $t = 130$),

where numerical solution $w_{130} = 5,487.020\ 363$

and analytical solution $R_{130} = 5,487.021\ 127$.

Maximum percentage relative error $= 3.411 \times 10^{-5}$ (at $t = 10$),

where numerical solution $w_{10} = 11.025\ 809$

and analytical solution $R_{10} = 11.025\ 812$.

As $t \to \infty$ we can again use equation (A1.12) to find $R_{\infty} = 10,509.017\ 808$, which compares well with $w_{800} = 10,509.017\ 814$.

Table A1.5. Comparison of results for the case $\beta = 9 \times 10^{-6}$, $\gamma = 0.85$.

t_i	Analytical solution R_i	Numerical solution w_i	Absolute error $\lvert R_i - w_i \rvert$	Percentage error $\lvert (R_i - w_i)/R_i \rvert \times 100\%$
0	0.000 000	0.000 000	0.000E+00	0.000E+00
10	11.025 812	11.025 809	3.761E−06	3.411E−05
20	29.181 540	29.181 530	9.726E−06	3.333E−05
30	59.023 965	59.023 946	1.933E−05	3.276E−05
40	107.930 739	107.930 704	3.458E−05	3.204E−05
50	187.693 297	187.693 238	5.923E−05	3.155E−05
60	316.756 242	316.756 144	9.858E−05	3.112E−05
70	522.948 046	522.947 890	1.567E−04	2.996E−05
80	845.749 543	845.749 890	2.419E−04	2.860E−05
90	1 335.406 515	1.335.406 158	3.568E−04	2.672E−05
100	2 043.712 930	2.043.712 425	5.049E−04	2.470E−05
110	3 000.950 359	3.000.949 711	6.474E−04	2.157E−05
120	4 182.271 905	4.182.271 158	7.474E−04	1.787E−05
130	5 487.021 127	5.487.020 363	7.635E−04	1.391E−05
140	6 762.745 079	6.762.744 389	6.902E−04	1.020E−05
150	7 869.794 649	7.869.794 091	5.577E−04	7.087E−06
160	8 735.290 774	8.735.290 364	4.101E−04	4.695E−06
170	9 358.368 904	9.358.368 624	2.808E−04	3.001E−06
180	9 780.782 644	9.780.782 456	1.882E−04	1.924E−06
190	10 055.623 755	10.055.623 635	1.201E−04	1.195E−06
200	10 229.695 021	10.229.694 944	7.672E−05	7.500E−07

The effect of variation of removal rate on the total number of removals is assessed by considering the results in Figure A1.3 for the cases $\gamma = 0.89$, $\gamma = 0.80$ and $\gamma = 0.85$ over a time period of 1200 days. Clearly the gradients of the three curves are different, with the lower the removal rate the steeper the curve. Also the lower the removal rate the shorter the time period required for the number of removals to level out. Clearly, as would be expected, smaller infection rates lead to higher values of R_∞.

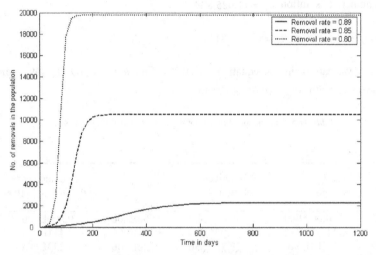

Figure A1.3. Plot of the number of removals from the population against time in days for different removal rates.

6. Interpretation and Conclusions

The case study has extended past work by Caldwell and Ram [3] to consider a deterministic model in the theory of contagious disease. A more realistic case has been considered by introducing a third variable, namely, the number of removals $R(t)$ at time t. Numerical results have been obtained by solving the resulting system of differential equations using the Runge-Kutta-Fehlberg method including error control to validate against analytical results. By comparing absolute and relative errors over long time periods there is clearly good agreement between analytical and numerical results. The Runge-Kutta-Fehlberg method, which requires only six functional evaluations per time step, is clearly an effective numerical tool and, in a similar way, could be applied to more complex models where analytical solutions are not possible. Tables of results and graphs have been presented to show the effects of variation of the infection rate and the removal rate on the number of removals from the population over long time periods.

7. Computer Algorithms

The Runge-Kutta-Fehlberg (R-K-F) method has been used to obtain the numerical results. The algorithm used is presented in Fehlberg [4] with more details given in Caldwell and Ram [3]. The important details are given in Appendices I and II.

8. References and Bibliography

1. Bailey, N.T.J., *The Mathematical Theory of Epidemics*, C.Griffin, London, 1957.
2. Bailey, N.T.J., *The Mathematical Approach to Biology and Medicine*, John Wiley & Sons, London, 1967.
3. Caldwell, J. and Ram, Y.M., "Study of a deterministic model in the theory of contagious disease," Int. J. Math. Educ. Sci. Techol., **26**, 1995, 639-646.
4. Fehlberg, E., "Klassische Runge-Kutta-Formeln vierter und niedrigerer Ordnung mit Schrittweiten-Kontrolle und ihre Anwendung auf Wärmeleitungsprobleme," Computing, **6**, 1970, 61-71.
5. Gerald, C.F. and Wheatley, P.O., *Applied Numerical Analysis*, 4th ed., Addison-Wesley, Reading, MA, 1989.
6. Maron, M.J., *Numerical Analysis. A Practical Approach*, Macmillan, New York, 1982.
7. Rice, J.R., *Numerical Methods, Software and Analysis*, Academic Press, Boston, 1993.

APPENDIX I – NUMERICAL SOLUTION BY RUNGE-KUTTA-FEHLBERG ALGORITHM

One popular technique for solving the initial value problem

$$\frac{dy}{dt} = f(t,y), \ a \le t \le b, \ y(a) = \alpha.$$ (A1.13)

is the Runge-Kutta-Fehlberg (R-K-F) method presented by Fehlberg [4] in 1970. This technique consists of using a Runge-Kutta method of order 5 $(\overline{w})$ to estimate the local truncation error in a Runge-Kutta method of order 4 (w). A clear advantage of this method is that it only requires six functional evaluations per step, whereas with arbitrary Runge-Kutta methods of orders 4 and 5 used together, we would require 10 functional evaluations per step. Error control theory (see Gerald and Wheatley [5], Maron [6] and Rice [7]) shows that, in the usual notation, the general error control inequality is given by

$$q \le \left\{ \frac{\varepsilon h}{|\overline{w}_i - w_i|} \right\}^{1/n}.$$ (A1.14)

For the R-K-F method of order 4(5) the usual choice of q is as follows:

$$q \le \left\{ \frac{\varepsilon h}{2|\overline{w}_i - w_i|} \right\}^{1/4} = 0.84 \left\{ \frac{\varepsilon h}{|\overline{w}_i - w_i|} \right\}^{1/4}.$$ (A1.15)

The algorithm below is based on the equations presented by Fehlberg [4] and the coefficients are given in Appendix II. The step size, which is variable, is denoted by h, the tolerance, which is the value the local truncation error is not to exceed, is denoted by TOL, and the number of iterations is denoted by the variable i.

Initialization $w_0 = \alpha$, $\overline{w} = \alpha$, $i = 0$, $h = (\text{TOL})^{1/4}$, $t = a$.
while $t \le b$
    $t := t + h$
    $i := i + 1$
    $w_i := w_{i-1} + \frac{25}{216} k1 + \frac{1408}{2565} k3 + \frac{2197}{4104} k4 - \frac{1}{5} k5$
    $\overline{w}_i := w_{i-1} + \frac{16}{135} k1 + \frac{6656}{12825} k3 + \frac{28561}{56430} k4 - \frac{9}{50} k5 + \frac{2}{55} k6$
    if $|\overline{w}_i - w_{i-1}| = 0$ then
        $h := h_{max}$

else

$$q := 0.84 \left\{ \frac{TOL}{|\overline{w}_i - w_i|} \right\}^{1/4}$$

if $gh < h_{min}$ then

    STOP—minimum step size exceeded

else

    if $qh < h$ then

        $t := t - h + qh$

        $h := qh$

$$w_i := w_{i-1} + \frac{25}{216}k1 + \frac{1408}{2565}k3 + \frac{2197}{4104}k4 - \frac{1}{5}k5$$

$$\overline{w}_i := w_{i-1} + \frac{16}{135}k1 + \frac{6656}{12825}k3 + \frac{28561}{56430}k4 - \frac{9}{50}k5 + \frac{2}{55}k6$$

    else

        $h := qh$

    endif

    endif

endif

endwhile

where,

$$k1 := h f[t, w_{i-1}]$$

$$k2 := h f\left[t + \frac{1}{4}h, w_{i-1} + \frac{1}{4}k1\right]$$

$$k3 := h f\left[t + \frac{3}{8}h, w_{i-1} + \frac{3}{32}k1 + \frac{9}{32}k2\right]$$

$$k4 := h f\left[t + \frac{12}{13}h, w_{i-1} + \frac{1932}{2197}k1 - \frac{7200}{2197}k2 + \frac{7296}{2197}k3\right]$$

$$k5 := h f\left[t + h, w_{i-1} + \frac{439}{216}k1 - 8k2 + \frac{3680}{513}k3 - \frac{845}{4104}k4\right]$$

$$k6 := h f\left[t + \frac{1}{2}h, w_{i-1} - \frac{8}{27}k1 + 2k2 - \frac{3544}{2565}k3 + \frac{1859}{4104}k4 - \frac{11}{40}k5\right]$$

The algorithm given above was combined into a MATLAB computer program which was used to solve the differential equation.

APPENDIX II – COEFFICIENTS FOR RUNGE-KUTTA-FEHLBERG METHOD

The following equations are presented by Fehlberg [4]:

$$f_0 = f(x_0, y_0)$$

$$f_x = f\left(x_0 + \alpha_x h, y_0 + h\sum_{\lambda=0}^{x-1}\beta_{x\lambda}f_\lambda\right), (x=1,2,3,4,5)$$

$$y = y_0 + h\sum_{x=0}^{4}c_x f_x + O(h^5) \cdot$$

$$\bar{y} = y_0 + h\sum_{x=0}^{5}\bar{c}_x f_x + O(h^6) \cdot$$

	λ	$\beta_{x\lambda}$					c_x	$\bar{c}_x$
x	α_x	0	1	2	3	4		
0	0	0					$\dfrac{25}{216}$	$\dfrac{16}{135}$
1	$\dfrac{1}{4}$	$\dfrac{1}{4}$					0	0
2	$\dfrac{3}{8}$	$\dfrac{3}{32}$	$\dfrac{9}{32}$				$\dfrac{1408}{2565}$	$\dfrac{6656}{12825}$
3	$\dfrac{12}{13}$	$\dfrac{1932}{2197}$	$-\dfrac{7200}{2197}$	$\dfrac{7296}{2197}$			$\dfrac{2197}{4104}$	$\dfrac{28561}{56430}$
4	1	$\dfrac{439}{216}$	-8	$\dfrac{3680}{513}$	$-\dfrac{845}{4104}$		$-\dfrac{1}{5}$	$-\dfrac{9}{50}$
5	$\dfrac{1}{2}$	$-\dfrac{8}{27}$	2	$-\dfrac{3544}{2565}$	$\dfrac{1859}{4104}$	$-\dfrac{11}{40}$		$\dfrac{2}{55}$

Case Study A2

ELECTROMAGNETIC FORCES IN HIGH FIELD MAGNET COILS

SUMMARY: A limiting design of large high field superconducting magnets is the problem of supporting the electromagnetic forces. It is therefore important to be able to estimate the forces appearing on the windings of magnets. A simple mathematical model is obtained which represents the stress distribution in magnet windings. In deriving the equations a number of simplifying assumptions have been made. This model is validated by checking the accuracy against homogeneous thick cylinder theory which involves the calculation of stresses by solving the Timoshenko stress equations. In this way, values of the circumferential stress have been compared for two separate coil configurations and conclusions are drawn.

1. Background

The problem of supporting the electromagnetic forces acting on the windings has becomes a limiting factor in the design of large high field superconducting magnets and has attracted much attention in work on superconducting motors and generators. The three important aspects to be considered are:

(i)   the effect of tensile stress on the performance of superconducting materials and composites;
(ii)   the calculation of the forces appearing on the windings of magnets;
(iii)   the experimental verification of these calculations.

This case study is concerned with the second of these aspects.

Two theoretical problems arise in the determination of the stress acting on magnet windings. The first of these, the calculation of the electromagnetic forces generating the stresses, is relatively simple in theory, but difficult and laborious to perform in practice. The second is the determination of the way in which these forces are accommodated by the windings and is less tractable, involving assumptions about the mechanical behaviour of the coil, e.g., the extent to which the winding may be considered as a solid body.

This case study involves the development of a mathematical model to represent the stress distribution in magnet windings. A number of simplifying approximations will be made in deriving the mathematical formulation.

2. Problem Statement

In this case study we develop a mathematical model to represent the stress distribution in magnet windings stating clearly any simplifying approximations. It may be assumed that the tensile (or "circumferential") stress T' in a circular turn (radius r) of wire carrying a current density J in a uniform magnetic field B parallel to the axis of the loop is given by $T' = BJr$.

We validate the model by comparing the results with those obtained from homogeneous thick cylinder theory which involves the calculation of stresses by solving the Timoshenko stress equations. In this way, we can compare the computed values of the circumferential stress for the following two coil configurations:

(1) Coil 1 (used in the prototype superconducting motor developed at International Research and Development Co. Ltd., and later tested by driving a cooling water pump at Fawley Power Station, Southampton, UK) has the following parameters:

Inner radius	$r_1 = 1.11\text{m}$
Outer radius	$r_1 = 1.29\text{m}$
Winding length	$2b = 0.55\text{m}$
Total Ampere turns	$NI = 3 \times 10^6 \text{A}$.

(2) Coil 2 (part of the magnet for the $7.0 \text{Wb}/\text{m}^2$ bubble chamber at the Rutherford High Energy Laboratory, UK) is a Helmholtz pair with the following parameters:

Inner radius	0.95m
Outer radius	1.70m
Total winding length (including gap)	2.30m
Length of gap between coils	0.15m
Ampere turns (per coil)	$1.02 \times 10^7 \text{A}$.

Finally, we validate the results, comment on the accuracy and draw conclusions.

3. Model Formulation

3.1 ELECTROMAGNETIC FORCES

It can readily be shown that the tensile (or 'hoop') stress T' in a circular turn (radius r) of wire carrying a current density J in a uniform magnetic field B parallel to the axis of the loop is given by $T' = BJr$. This 'generated' force acts in fact only on the superconducting portion of a composite superconductor, since this alone is carrying current. It is possible that when the stresses are high the superconductor may cut through the copper, but we assume for the present purpose that this does not occur. If the bond between the superconductor and copper is rigid, then some of the stress will be transmitted to the copper as the two materials must strain by the same amount, and in this case the stresses in the two materials will be in proportion to their Young's moduli. For copper and niobium-titanium these are approximately equal, though copper has a very much lower yield point, and so we assume that the stress is uniform across the composite conductor.

In the central plane of a superconducting coil the electromagnetic force is a pure radial bursting force (giving rise to the tensile stress) but off this plane there are axial components of force since the magnetic field is not in general parallel to the coil axis. These forces, though sometimes large, are ignored for the purposes of the calculations below.

3.2 STRESS ON THE WINDINGS

Consider a winding of inner and outer radii r_1 and r_2, respectively, carrying a uniform current density J. We assume that a disc element of unit length in the central plane can be considered as a solid body, with zero radial compressibility and uniform radial deflection δ due to the electromagnetic forces acting on it. As stated above, we ignore axial forces transmitted from windings off the mid-plane. Then the strain at radius r is δ/r and hence the stress T in the windings at radius r is given by:

$$E = \frac{T}{\delta/r} \qquad (A2.1)$$

i.e., $Tr = E\delta$, where E is Young's modulus for the material.

The tensile force on the cross-section of an element dr thick is therefore:

$$Tdr = E\delta\frac{dr}{r} \qquad (A2.2)$$

and the total tensile force on the disc is:

$$F = \int_{r_1}^{r_2} T \, dr = E\delta \int_{r_1}^{r_2} \frac{dr}{r} = E\delta \ln(r_2 / r_1) \cdot \tag{A2.3}$$

Now if we further assume that the coil was wound under zero tension (i.e., there are no forces on the winding former) and there is no external hydrostatic pressure difference across the windings then the force on the windings is purely electromagnetic in origin, ignoring the weight of the windings themselves. But the tensile force generated in the element dr is $BJr\,dr$, and therefore:

$$F = \int_{r_1}^{r_2} BJr \, dr \cdot \tag{A2.4}$$

From the three equations (A2.2), (A2.3) and (A2.4) we obtain:

$$E\delta = Tr = \frac{\int_{r_1}^{r_2} BJr \, dr}{r \ln(r_2 / r_1)} \tag{A2.5}$$

and hence:

$$T = \frac{\int_{r_1}^{r_2} BJr \, dr}{r \ln(r_2 / r_1)} \cdot \tag{A2.6}$$

This is the tensile stress in the windings at radius r. It is worth noting again all the approximations made in arriving at this result, since they indicate the validity of its application:

(i)   The bond between the two materials is rigid.
(ii)   The stress in the two materials is the same (i.e., they have the same Young's modulus).
(iii)   Axial forces (and hence the introduction of Poisson's ratio as a parameter) can be ignored.
(iv)   The coil acts as a solid body, i.e., the radial deflection under the stress is uniform.
(v)   The forces acting on the coil are purely electromagnetic in origin.

Computer programs are available for the calculation of the flux density within the windings of solenoids. One such program developed by Culwick [3] is available from the Brookhaven National Laboratory. This can be modified to obtain the expression:

$$\int_{r_1}^{r_2} BJr \, dr \cdot$$

The electromagnetic forces cannot be determined by hand using this method, as the calculation involves successive approximations with many iterations. It is based on the elliptic integral formulation of the field around a single turn of wire. An alternative method which can be used by hand, though laboriously, uses the relationship between electromagnetic 'pressure' at a point and the mutual inductance between the coil and a single turn passing through that point.

The particular type of coil of most interest to engineering applications associated with superconducting motors and generators is that which has a large bore but relatively small winding cross-section. For this type of coil the flux density varies approximately linearly with radius across the windings except around the zero field region where Bdr is small. This means that if $B(r_1)$ and $B(r_2)$ are known, the integral can be evaluated directly. Using the linear relationship we can write:

$$B(r) = B(r_1) + \frac{r - r_1}{r_2 - r_1}(B(r_2) - B(r_1)) \tag{A2.7}$$

and then equation (A2.6) simplifies to give:

$$T = \frac{J\{B_1(r_2^2 + r_1 r_2 - 2r_1^2) + B_2(2r_2^2 - r_1 r_2 - r_1^2)\}}{6r \ln(r_2 / r_1)}, \tag{A2.8}$$

where $B_1 = B(r_1)$ and $B_2 = B(r_2)$.

Alternatively, we can make the further approximation:

$$\int_{r_1}^{r_2} BJr \, dr = Jr_m \int_{r_1}^{r_2} B \, dr$$

$$= \tfrac{1}{2} Jr_m (r_2 - r_1)(B_1 + B_2), \tag{A2.9}$$

where $r_m = \tfrac{1}{2}(r_1 + r_2)$ is the mean radius of the coil. In this case equation (A2.6) simplifies to give:

$$T = \frac{J(B_1 + B_2)(r_2^2 - r_1^2)}{4r \ln(r_2 / r_1)}. \tag{A2.10}$$

For the types of coil we are considering $r_1 \approx r_2$ and we can say:

$$r \ln(r_2 / r_1) \approx r_2 - r_1,$$

which gives a constant stress throughout the windings of

$$T = \tfrac{1}{4} J (B_1 + B_2)(r_1 + r_2). \tag{A2.11}$$

This equation (A2.11) is simple to use as B_1 and B_2 can be obtained from curves plotting these factors as a function of the winding configuration.

Equations (A2.8), (A2.10) and (A2.11) give the tensile stress with decreasing accuracy.

3.3 VALIDITY OF THE EQUATIONS

Because of the drastic simplifications made in the above calculations, the expressions obtained must be applied with caution. It would appear that they are least inaccurate for the case of a 'pancake' winding in the mid-plane of a solenoid wound without interlayer spacers with a conductor of large cross-section. If a packing factor must be taken into account, the current density must clearly be averaged over the solid load-bearing cross-section.

The results given above are intended only as a starting point for a more detailed and accurate consideration of the stresses appearing on magnet windings in different circumstances. A more accurate calculation of the stresses can be made by solving the Timoshenko stress equations and this has been done by Middleton and Trowbridge [5] to produce the formula:

$$\sigma_\theta = \sigma_\theta(r_1) - v\sigma_z(r_1) - \sigma_r + v\sigma_z - (1+v)\int_{r_1}^{r_2} R\,dr, \tag{A2.12}$$

where σ_z, σ_r, σ_θ are the three stress components, v is Poisson's ratio and $R = B_z J$ is the radial component of the electromagnetic body force. The tensile (or circumferential) stress σ_θ is equivalent to T in this notation.

4. Mathematical Solution

4.1 HOMOGENEOUS THICK CYLINDER THEORY

The electromagnetic body forces for an assembly of solenoids are calculated by a computer program developed by Culwick [3] which is available from Brookhaven National Laboratory. This program divides each solenoid into a number of uniformly spaced filamentary conductors. Each filament is then repeatedly subdivided by 4 until the sum of the fields of the subdivisions is equal to the field of the filament within a prescribed tolerance. The forces are then calculated from the equations:

$$R = B_z J, \tag{A2.13}$$

$$Z = B_r J, \qquad (A2.14)$$

where J is the current density and R, Z are the radial and axial components of force per unit volume and B_r, B_z are the corresponding field components.

The equations relating the stress components and strains for axially symmetric systems when shear is neglected are:

$$\frac{d\sigma_r}{dr} + \frac{(\sigma_r - \sigma_\theta)}{r} + R(r,z) = 0 \qquad (A2.15)$$

$$\frac{d\sigma_z}{dz} + Z(r,z) = 0 \qquad (A2.16)$$

$$Ee_r = \sigma_r - v\sigma_\theta - v\sigma_z \qquad (A2.17)$$

$$Ee_\theta = -v\sigma_r + \sigma_\theta - v\sigma_z \qquad (A2.18)$$

$$Ee_z = -v\sigma_r - v\sigma_\theta + \sigma_z, \qquad (A2.19)$$

where σ_r, σ_θ and σ_z are the radial, circumferential and axial stresses, respectively, and e_r, e_θ and e_z the corresponding strains. E is Young's modulus and v is Poisson's ratio for the material.

By introducing:

$$e_r = \frac{du}{dr}, \; e_\theta = \frac{u}{r}, \qquad (A2.20)$$

where u is the radial displacement, it is possible to solve equations (A2.15)-(A2.19) and thus find an expression for the circumferential stress σ_θ. This has been done by Middleton and Trowbridge [5] but, since they only outline the method without including a detailed derivation of the equations, this has been included below.

On eliminating u from equation (A2.20) we obtain:

$$r\frac{de_\theta}{dr} = -e_\theta + e_r. \qquad (A2.21)$$

From equations (A2.17) and (A2.18) we have:

$$E(e_r - e_\theta) = (1 + v)(\sigma_r - \sigma_\theta).$$ (A2.22)

On differentiating equation (A2.18) with respect to r we obtain:

$$E\frac{de_\theta}{dr} = -v\frac{d\sigma_r}{dr} + \frac{d\sigma_\theta}{dr} - v\frac{d\sigma_z}{dr},$$ (A2.23)

and using this equation with equation (A2.21) gives:

$$Er\frac{de_\theta}{dr} = E(e_r - e_\theta) = r\left(\frac{d\sigma_\theta}{dr} - v\frac{d\sigma_r}{dr} - v\frac{d\sigma_z}{dr}\right).$$

From equations (A2.15) and (A2.22) we have:

$$\frac{(1+v)(\sigma_r - \sigma_\theta)}{r} = \frac{E}{r}(e_r - e_\theta)$$

$$= \frac{d\sigma_\theta}{dr} - v\frac{d\sigma_r}{dr} - v\frac{d\sigma_z}{dr}$$

$$= -(1+v)\left(\frac{d\sigma_r}{dr} + R\right)$$

and hence:

$$\left(\frac{d\sigma_\theta}{dr} + \frac{d\sigma_r}{dr}\right) - v\frac{d\sigma_z}{dr} + (1+v)R = 0.$$ (A2.24)

On integrating equation (A2.24) with respect to r we obtain:

$$\sigma_\theta + \sigma_r = (\sigma_{\theta 1} - v\sigma_{z 1}) + v\sigma_z(r, z) - (1+v)\int_{r_1}^{r} R\,dr,$$

where $\sigma_{\theta 1}$ and $\sigma_{z 1}$ are the circumferential and axial stresses at the inner coil radius $r = r_1$ for some z plane. The radial stresses at the inner ($r = r_1$) and outer radii ($r = r_2$) are taken as zero in this case.

Using the notation:

$$\int_{r_1}^{r} r^n R\,dr = I_n,$$

the sum of the circumferential and radial stresses becomes:

$$(\sigma_\theta + \sigma_r) = (\sigma_{\theta 1} - v\sigma_{z1}) + v\sigma_z - pI_0, \qquad \text{(A2.25)}$$

where $p = 1 + v$.

Rewriting equation (A2.15) as:

$$\sigma_r - \sigma_\theta = -rR - r\frac{d\sigma_r}{dr},$$

and adding to equation (A2.25) yields:

$$2\sigma_r = (\sigma_{\theta 1} - v\sigma_{z1}) + v\sigma_z - pI_0 - rR - r\frac{d\sigma_r}{dr},$$

which, on rearranging and multiplying by r, gives:

$$\frac{d}{dr}(r^2\sigma_r) = (\sigma_{\theta 1} - v\sigma_{z1}) + vr\sigma_z - pI_0 r - r^2 R \cdot$$

Integration with respect to r yields:

$$r^2\sigma_r = \tfrac{1}{2}(\sigma_{\theta 1} - v\sigma_{z1})(r^2 - r_1^2)$$
$$+ v\int_{r_1}^r r\sigma_z \, dr - p\int_{r_1}^r rI_0 \, dr - I_2.$$

The third term on the right-hand-side can be integrated by parts to give:

$$\int_{r_1}^r rI_0(r) \, dr = \frac{r^2}{2}I_0 - \tfrac{1}{2}I_2$$

and hence the radial stress is given by:

$$\sigma_r = \frac{1}{2}(\sigma_{\theta 1} - v\sigma_{z1})\left(1 - \frac{r_1^2}{r^2}\right) - \frac{p}{2}I_0 - \frac{1}{r^2}\left(1 - \frac{p}{2}\right)I_2 + \frac{v}{r^2}\int_{r_1}^r r\sigma_z \, dr \cdot \quad \text{(A2.26)}$$

Note that if the axial stress σ_z is constant, then:

$$\frac{v}{r^2}\int_{r_1}^r r\sigma_z \, dr = \frac{v\sigma_{z1}}{2}\left(1 - \frac{r_1^2}{r^2}\right)$$

and therefore:

$$\sigma_r \rightarrow \frac{\sigma_{\theta 1}}{2}\left(1-\frac{r_1^2}{r^2}\right) - \frac{p}{2}I_0 - \frac{1}{r^2}\left(1-\frac{p}{2}\right)I_2 .$$

Clearly the axial stress σ_z is obtained by direct integration of equation (A2.16) and $\sigma_{\theta 1}$ is obtained from equation (A2.26) by inserting the condition that $\sigma_r = 0$ at $r = r_2$.

Hence the sequence of computation is as follows:

(1) σ_z from equation (A2.16).
(2) $\sigma_{\theta 1}$ from equation (A2.26) with $r = r_2$.
(3) σ_r from equation (A2.26).
(4) σ_θ from equation (A2.25).
(5) e_r, e_θ, e_z from equation (A2.17)-(A2.19).
(6) u from equation (A2.20).

From equation (A2.25) the circumferential stress is given by:

$$\sigma_\theta = \sigma_\theta(r_1) - v\sigma_z(r_1) - \sigma_r + v\sigma_z - (1+v)\int_{r_1}^{r_2} R\, dr \qquad (A2.27)$$

and this expression, together with the other stress and strain components, is then evaluated using a computer program.

5. Model Validation

By making a number of simplifying assumption Caldwell [2] obtained the following three equations for the circumferential stress:

$$\sigma_\theta = \frac{J\{B_1(r_2^2 + r_1 r_2 - 2r_1^2) + B_2(2r_2^2 - r_1 r_2 - r_1^2)\}}{6r\,\ln(r_2/r_1)}, \qquad (A2.28)$$

$$\sigma_\theta = \frac{J\{(B_1 + B_2)(r_2^2 - r_1^2)\}}{4r\,\ln(r_2/r_1)}, \qquad (A2.29)$$

$$\sigma_\theta = \tfrac{1}{4}J(B_1 + B_2)(r_1 + r_2), \qquad (A2.30)$$

where $B_1 = B(r_1)$ and $B_2 = B(r_2)$.

In order to assess the accuracy of these equations, values of the circumferential stress σ_θ (which is the dominant one) have been calculated for each of the equations (A2.28), (A2.29) and (A2.30) and compared with the values obtained by the more rigorous method (equation (A2.27)) for two different coil configurations.

Coil 1 was used in the prototype superconducting motor developed at International Research and Development Co. Ltd., and later tested by driving a cooling water pump at Fawley Power Station and had parameters as given in Section 2. For equation (A2.28), (A2.29) and (A2.30), the values of B_1 and B_2 were obtained from the BNL program [1]. The circumferential stresses obtained including those from equation (A2.27) are plotted in Figure A2.1.

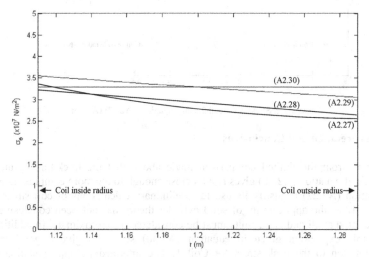

Figure A2.1. Stresses for coil 1 using equations (A2.27)-(A2.30).

The second coil magnet was for the $7.0\,\mathrm{Wb/m^2}$ bubble chamber at the Rutherford High Energy Laboratory. The solution of equation (A2.27) for this coil is given by Middleton and Trowbridge [5] for a particular value of z and the variation of $B_z(r)$ with radius is also plotted. Values of B_1 and B_2 were obtained from this plot. Coil 2 is a Helmholtz pair with parameters as given in Section 2. The circumferential stresses for the coil obtained by using equations (A2.27)-(A2.30) are plotted in Figure A2.2.

On comparing the results we find that in both cases the discrepancy between equations (A2.26) and (A2.27), though significant, is not large and that for both coils equation (A2.29) gives the highest value of stress. At $r = r_1$, where the peak stress occurs, equations (A2.28) and (A2.27) are in close agreement for Coil 1 but the agreement is not particularly good for Coil 2.

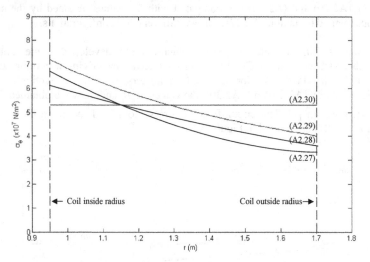

Figure A2.2. Stresses for coil 2 using equations (A2.27)-(A2.30).

6. Interpretation and Conclusions

It appears from the limited comparison made above that the thick/thin cylinder approach of equation (A2.28) gives results close enough to the more complex theory of equation (A2.27) to justify its use for preliminary calculations of coil stresses, particularly as the applicability of solid cylinder theory has not been conclusively demonstrated at all and any results must be considered as approximate. In addition, the simple expression given in equation (A2.30) seems to give a good first approximation to the peak stress for Coil 1. The intermediate approximation of equation (A2.29) is apparently of little value except that when used it seems to err on the high side which is useful information. None of the equations (A2.28), (A2.29) or (A2.30) can, however, be a complete substitute for the full theory, if only because they give no information about the radial and axial stresses.

The comparison has been carried out for two particular coil geometries which have been used in practice. However, comparisons made for other theoretical coil geometries provide similar agreement and this effectively gives us more confidence in the simplified formulae.

7. Computer Algorithms

Computer algorithms are available for the calculation of the flux density B within the windings of solenoids. One such algorithm developed by Culwick [3], which is available from the Brookhaven National Laboratory, was used to calculate the electromagnetic body forces for an assembly of solenoids. Essentially this program divides each solenoid into a number of uniformly spaced filamentary conductors. Each filament is then repeatedly subdivided by 4 until the sum of the fields of the subdivisions is equal to the fields of the filament within a prescribed tolerance.

8. References and Bibliography

1. Appleton, A.D., Cowhig, T.P. and Caldwell, J., "Some design aspects of a large superconducting magnet," Proc. 2nd Int. Conf. Magnet Technology, Oxford, 1967, 553-559.
2. Caldwell, J., "The circumferential stress in large high field magnet coils," Appl. Math. Modelling, **4**, 1980, 228-229.
3. Culwick, B.B., Brookhaven National Laboratory Report No. HH05-0, BBC-1, 1965 (unpublished).
4. Melville, D. and Mattocks, P.G., "Stress calculations for high magnetic field coils," J. Phys. D: Appl. Phys., **5**, 1972, 1745-1759.
5. Middleton, A.J. and Trowbridge, C.W., "Mechanical stresses in large high field magnet coils," Proc. 2nd Int. Conf. Magnet Technology, Oxford, 1967, 140.
6. Mulhall, B.E. and Prothero, D.H., "Mechanical stresses in solenoid coils," J. Phys. D: Appl. Phys., **6**, 1973, 1973-1977.
7. Timoshenko, S. and Goodier, J.N., *Theory of Elasticity*, McGraw-Hill, London, 1951.

Project B1

MASS BALANCE OF A REACTOR IN STEADY STATE

SUMMARY: The design of a chemical reactor is particularly important in the field of chemical engineering. The design of the system allows the chemical reaction to take place in a safe and efficient way. A mathematical model is built which represents the concentration of the chemical along the reactor in the steady state case. Both numerical and analytical solutions will be presented and compared for accuracy purposes. This kind of study can apply to real life problems, such as waste treatment.

1. Background

The conservation of mass has been an active research area in the field of chemical engineering. The conservation of mass can be expressed as a balance of a chemical that enters and leaves a system volume, i.e., the volume enclosed by the boundaries. As time goes on, this idea can be represented as

$$
\begin{bmatrix} \text{rate of accumulation} \\ \text{of chemical} \\ \text{inside the system} \\ \text{(moles/time)} \end{bmatrix} = \begin{bmatrix} \text{rate of flow of} \\ \text{chemical} \\ \text{into the system} \\ \text{(moles/time)} \end{bmatrix} - \begin{bmatrix} \text{rate of flow of} \\ \text{chemical out} \\ \text{of the system} \\ \text{(moles/time)} \end{bmatrix} + \begin{bmatrix} \text{rate of generation} \\ \text{of chemical} \\ \text{inside the system} \\ \text{(moles/time)} \end{bmatrix}.
$$

We assume the chemical is neither created nor destroyed in the chemical reactor, because we have already employed the idea of conservation of mass. If the rate of entering of chemical is greater than the rate of leaving the system, the mass increases, or vice versa. If the chemical enters and leaves the system at the same rate, the accumulation of chemical inside the system will be zero. As time proceeds (until the stable condition is reached), this can be represented as

Flow in = Flow out.

Chemical engineers employ the conservation of mass to determine the steady-state concentration of a system of coupled reactors by expressing the inputs and outputs in terms of measurable variables and parameters. In this way it is possible to develop a mass balance and derive a differential equation for concentration.

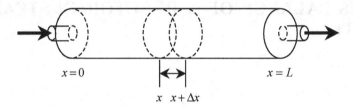

Figure B1.1. An elongated cylindrical reactor with a single entry and exit point.

2. Problem Statement

Figure B1.1 shows a cylindrical reactor with a single entry and exit point. The reactor can be characterized as a *distributed-parameter system*. If it is assumed that the chemical being modelled is subject to first-order decay and the cylindrical tube is well-mixed vertically and laterally, a mass balance can be performed on a finite segment of length Δx, as in

$$V\frac{\Delta c}{\Delta t} = \underbrace{Fc(x)}_{\text{Flow in}} - \underbrace{F\left[c(x) + \frac{\partial c(x)}{\partial x}\Delta x\right]}_{\text{Flow out}} - \underbrace{DA_c\frac{\partial c(x)}{\partial x}}_{\text{Dispersion in}}$$

$$+ \underbrace{DA_c\left[\frac{\partial c(x)}{\partial x} + \frac{\partial}{\partial x}\frac{\partial c(x)}{\partial x}\Delta x\right]}_{\text{Dispersion out}} - \underbrace{\gamma Vc}_{\text{Decay reaction}} , \qquad (B1.1)$$

where V = volume (m^3), F = flow rate (m^3/h), c is concentration (moles/m^3), D is a dispersion coefficient (m^2/h), A_c is the tank's cross-sectional area (m^2), and γ is the first-order decay coefficient (h^{-1}). Note that the dispersion terms are based on *Fick's first law*,

$$J = \text{Flux (moles/}hm^4) = -D\frac{\partial c}{\partial x}, \qquad (B1.2)$$

which is directly analogous to Fourier's law for heat conduction. It specifies that turbulent mixing tends to move mass from regions of high to low concentration. The parameter D, therefore, reflects the magnitude of turbulent mixing.

Use the above mass balance around a finite segment along the longitudinal axis of the cylindrical tank shown in Figure B1.1 to formulate a parabolic PDE. State clearly any modelling assumptions made in the formulation. Then consider the steady state solution of this equation.

First of all, find an analytical solution of the ODE boundary-value problem. Then find a numerical solution using finite differences and compare with the exact solution for certain assumed values of the key parameters. Determine how the concentration of the chemical varies with distance along the longitudinal axis of the cylindrical reactor for a chemical which decays with first-order decay kinetics.

Finally, validate the model by using the finite element method and draw conclusions.

3. Model Formulation

If Δx and Δt are allowed to approach zero, equation (B1.1) becomes

$$\frac{\partial c}{\partial t} = D\frac{\partial^2 c}{\partial x^2} - U\frac{\partial c}{\partial x} - \gamma c, \qquad (B1.3)$$

where $U = F/A_c$ is the velocity of the water flowing through the tank. The mass balance for Figure B1.1 is, therefore, now expressed as a parabolic partial differential equation. This equation (B1.3) is sometimes referred to as the advection-dispersion equation with first-order reaction. At steady state, it is reduced to a second-order ODE,

$$D\frac{d^2 c}{dx^2} - U\frac{dc}{dx} - \gamma c = 0. \qquad (B1.4)$$

Prior to $t = 0$, the reactor is filled with water which contains no chemical. Starting from $t = 0$, the chemical is injected into the reactor's inflow at a constant level of c_{in}. Thus, the following boundary conditions hold:

$$\begin{cases} Fc_{in} = Fc_0 - DA_c \dfrac{dc_0}{dx}, \\ c'(L,t) = 0. \end{cases} \qquad (B1.5)$$

The second condition specifies that chemical leaves the reactor purely as a function of flow through the outlet pipe. This means that it is assumed that dispersion in the reactor does not affect the exit rate. Under these conditions, we use numerical methods to solve equation (B1.4) for the steady-state levels in the reactor. Note that this is an ODE boundary-value problem.

4. Mathematical/Numerical Solution

4.1 ANALYTICAL SOLUTION

To formulate a mathematical model we make the following assumptions:

A1. Chemical being modelled is subject to first order decay.
A2. The tank is well-mixed vertically and laterally.
A3. Dispersion in the reactor does not affect the exit rate.

At steady state, we have a second order ODE boundary-value problem.

$$\begin{cases} Dc''(x) - Uc'(x) - \gamma c = 0, \ 0 < x < L, \\ Fc_{in} = Fc(0) - DA_c c'(0), \\ c'(L) = 0. \end{cases} \tag{B1.6}$$

Putting $A_c = F/U$, the first boundary condition can be further simplified and equation (B1.6) becomes

$$\begin{cases} Dc''(x) - Uc'(x) - \gamma c = 0, \ 0 < x < L, \\ c_{in} = c(0) - \dfrac{D}{U} c'(0), \\ c'(L) = 0. \end{cases} \tag{B1.7}$$

To obtain the solution analytically, we assume the solution takes the form $c(x) = e^{rx}$, resulting in the characteristic equation

$$Dr^2 - Ur - \gamma = 0. \tag{B1.8}$$

The discriminant $\Delta = U^2 + 4D\gamma$ is always positive, the roots are real and the general solution can be represented as

$$c(x) = A_1 e^{r_1 x} + A_2 e^{r_2 x}, \tag{B1.9}$$

where A_1 and A_2 are constants that can be determined from the initial conditions.

With the help of mathematical solver MAPLE, we obtain

$$A_1 = \frac{\%2 e^{\frac{\%2L}{2D}}}{\%3} U c_{in},$$
(B1.10)

$$A_2 = -\frac{\%1 e^{\frac{\%1L}{2D}}}{\%3} U c_{in},$$
(B1.11)

$$r_1 = \frac{\%1}{2D},$$
(B1.12)

$$r_2 = \frac{\%2}{2D},$$
(B1.13)

$$\%1 = U + \sqrt{U + 4D\gamma},$$
(B1.14)

$$\%2 = U - \sqrt{U + 4D\gamma},$$
(B1.15)

$$\%3 = e^{\frac{\%2L}{2D}} U^2 - e^{\frac{\%2L}{2D}} U \sqrt{U + 4D\gamma} + 2e^{\frac{\%2L}{2D}} D\gamma$$
$$- e^{\frac{\%1L}{2D}} U^2 - e^{\frac{\%1L}{2D}} U \sqrt{U + 4D\gamma} - 2e^{\frac{\%1L}{2D}} D\gamma.$$
(B1.16)

The mass fluxes for the steady-state solution can be computed using *Fick's first law*. Once we have the analytical form of $c(x)$, we can work out the analytical solution of mass fluxes by taking the first derivative of $c(x)$ with respect to x and multiplying by $-D$. This yields

$$J = -D \frac{dc}{dx} = -D(A_1 r_1 e^{r_1 x} + A_2 r_2 e^{r_2 x}).$$
(B1.17)

4.2 NUMERICAL SOLUTION

One more assumption is made before carrying out the numerical approximation, namely,

A4. At the reactor's ends, this process introduces nodes that lie outside the system, i.e., c_{-1} and c_{n+1} are introduced.

A steady state solution can be developed by substituting centered finite differences for the first and the second derivatives to give

$$D\frac{c_{i+1}-2c_i+c_{i-1}}{(\Delta x)^2} - U\frac{c_{i+1}-c_{i-1}}{2\Delta x} - \gamma c_i = 0 \cdot \tag{B1.18}$$

Rearranging terms gives

$$-\left(\frac{1}{2}+\frac{D'}{U\Delta x}\right)c_{i-1} + \left(\frac{\gamma\Delta x}{U}+\frac{2D}{U\Delta x}\right)c_i - \left(-\frac{1}{2}+\frac{D}{U\Delta x}\right)c_{i+1} = 0 \cdot \tag{B1.19}$$

This equation can be written for each of the system's nodes. At the reactor's ends, this process introduces nodes that lie outside the system. For example, at the inlet node ($i = 0$),

$$-\left(\frac{1}{2}+\frac{D}{U\Delta x}\right)c_{-1} + \left(\frac{\gamma\Delta x}{U}+\frac{2D}{U\Delta x}\right)c_0 - \left(-\frac{1}{2}+\frac{D}{U\Delta x}\right)c_1 = 0 \cdot \tag{B1.20}$$

The c_{-1} can be removed by imposing the first boundary condition. At the inlet, the following mass balance must hold:

$$Fc_{in} = Fc_0 - DA_c\frac{dc_0}{dx}, \tag{B1.21}$$

where c_0 represents the concentration at $x = 0$. Thus, this boundary condition specifies that the amount of chemical carried into the tank by advection through the pipe must be equal to the amount carried away from the inlet by both advection and turbulent dispersion in the tank. A finite divided difference can be substituted for the derivative

$$Fc_{in} = Fc_0 - DA_c\frac{c_1-c_{-1}}{2\Delta x}, \tag{B1.22}$$

which can be solved for

$$c_{-1} = c_1 + \frac{2\Delta xU}{D}c_{in} - \frac{2\Delta xU}{D}c_0, \tag{B1.23}$$

which can be substituted into (B1.20) to give

$$\left(\frac{\gamma\Delta x}{U}+\frac{2D}{U\Delta x}+\frac{\Delta xU}{D}+2\right)c_0 - \left(\frac{2D}{U\Delta x}\right)c_1 = \left(\frac{\Delta xU}{D}+2\right)c_{in} \cdot \tag{B1.24}$$

A similar exercise can be performed for the outlet, where the original difference equation is

$$-\left(\frac{1}{2}+\frac{D}{U\Delta x}\right)c_{n-1}+\left(\frac{\gamma\Delta x}{U}+\frac{2D}{U\Delta x}\right)c_n-\left(-\frac{1}{2}+\frac{D}{U\Delta x}\right)c_{n+1}=0. \qquad (B1.25)$$

The boundary condition at the outlet is

$$Fc_n-DA_c\frac{dc_n}{dx}=Fc_n. \qquad (B1.26)$$

As with the inlet, a divided difference can be used to approximate the derivative,

$$Fc_n-DA_c\frac{c_{n+1}-c_{n-1}}{2\Delta x}=Fc_n. \qquad (B1.27)$$

Solving the equation (B1.27) gives $c_{n+1}=c_{n-1}$. In other words, the slope at the outlet must be zero for equation (B1.27) to hold. Substituting this result into (B1.25) and simplifying gives

$$-\left(\frac{2D}{U\Delta x}\right)c_{n-1}+\left(\frac{\gamma\Delta x}{U}+\frac{2D}{U\Delta x}\right)c_n=0. \qquad (B1.28)$$

Equations (B1.19), (B1.24) and (B1.28) now form a system of n tridiagonal equations with n unknowns.

For example, if $D=1$, $U=1$, $\Delta x=2.5$, $\gamma=0.2$, $c_{in}=100$ and $L=10$, the system is

$$\begin{bmatrix} 5.8 & -0.8 & & & \\ -0.9 & 1.3 & 0.1 & & \\ & -0.9 & 1.3 & 0.1 & \\ & & -0.9 & 1.3 & 0.1 \\ & & & -0.8 & 1.3 \end{bmatrix}\begin{bmatrix} c_0 \\ c_1 \\ c_2 \\ c_3 \\ c_4 \end{bmatrix}=\begin{bmatrix} 450 \\ 0 \\ 0 \\ 0 \\ 0 \end{bmatrix},$$

which can be solved to give $c_0=85.34$, $c_1=56.23$, $c_2=37.05$, $c_3=24.49$, $c_4=15.07$.

Table B1.1 shows the concentration of the reactor at $x=0$, 2.5, 5, 7.5 and 10 m respectively.

Table B1.1. Concentration of the reactor at $x = 0, 2.5, 5, 7.5, 10$m.

c (moles / m^3)	$x = 0$m	$x = 2.5$m	$x = 5$m	$x = 7.5$m	$x = 10$m
Exact Solution	85.4102	55.7245	36.3626	23.8408	17.7334
$\Delta x = 0.125$s	85.4100	55.7259	36.3644	23.8417	17.7247
$\Delta x = 0.25$s	85.4095	55.7299	36.3697	23.8443	17.6987
$\Delta x = 2.5$s	85.3426	56.2335	37.0471	24.4888	15.0700

These results (analytical and numerical) are plotted in Figure B1.2. As expected, the concentration decreases due to the decay reaction as the chemical flows through the tank. In addition to the above computation, Figure B1.2 shows another case with $D = 4$. Notice how increasing the turbulent mixing tends to flatten the curve.

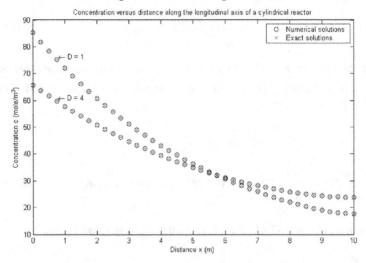

Figure B1.2a. Concentration versus distance along the longitudinal axis of a cylindrical reactor for a chemical that decays with first-order decay kinetics with $\Delta x = 0.25$.

5. Model Validation

We attempt to solve the equation (B1.4) by the finite element method. Equation (B1.4) can be written as

$$D \frac{d^2 c}{dx^2} - U \frac{dc}{dx} = \gamma c. \qquad (B1.29)$$

Dividing equation (B1.29) by D gives

$$\frac{d^2c}{dx^2} - \frac{U}{D}\frac{dc}{dx} = \frac{\gamma}{D}c \,. \tag{B1.30}$$

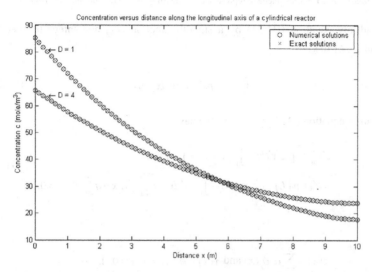

Figure B1.2b. Concentration versus distance along the longitudinal axis of a cylindrical reactor for a chemical that decays with first-order decay kinetics with $\Delta x = 0.125$.

Let $p = -U/D$, $q = -\gamma/D$, then equation (B1.30) becomes

$$c'' + pc' + qc = 0 \,, \tag{B1.31}$$

subject to boundary conditions

$$\begin{cases} c_{in} = c(0) - \dfrac{D}{U}c'(0), \\ c'(L) = 0. \end{cases} \tag{B1.32}$$

The boundary conditions give

$$c'(L) = 0, \ c'(0) = -p(c_0 - c_{in}) \,. \tag{B1.33}$$

Now we consider the discretization. Since the tank is assumed well-mixed vertically and laterally, therefore it is just a one-dimensional problem. We divide the tank into N elements of length h where $Nh = L = 10$. Thus, the system consists of N elements and $N+1$ nodes (see Figure B1.3).

Figure B1.3. The finite-element representation consisting of N elements and $N+1$ nodes.

We multiply equation (B1.31) by a smooth function $v(x)$ and integrate over the interval $[0, L]$ to give

$$\int_0^L (c'' + pc' + qc)v\,dx = 0. \tag{B1.34}$$

Integrating equation (B1.34) by parts, we have

$$c'v\Big|_0^L - \int_0^L c'v'\,dx + \int_0^L pc'v\,dx + \int_0^L qcv\,dx$$
$$= c'(L)v(L) - c'(0)v(0) - \int_0^L c'v'\,dx + p\int_0^L c'v\,dx + q\int_0^L cv\,dx = 0. \tag{B1.35}$$

Let

$$c(x) = \sum_{j=0}^N \alpha_j \phi_j(x) \text{ and } v(x) = \phi_i(x) \text{ for } i = 0, 1, \cdots, N,$$

and

$$c_i \approx c(x_i), \tag{B1.36}$$

where

$$\phi_i(x_j) = \begin{cases} 0, & \text{if } i \neq j, \\ 1, & \text{if } i = j. \end{cases}$$

Therefore, equation (B1.35) can be written as

$$-\sum_{j=0}^N \alpha_j \phi_j'(0)\phi_i(0) - \sum_{j=0}^N \alpha_j \int_0^L \phi_i'\phi_j'\,dx$$
$$+ p\sum_{j=0}^N \alpha_j \int_0^L \phi_i\phi_j'\,dx + q\sum_{j=0}^N \alpha_j \int_0^L \phi_i\phi_j\,dx = 0, \tag{B1.37}$$

which can be solved for

$$p(c_0 - c_{in})\phi_i(0) - \sum_{j=0}^{N} \alpha_j \int_0^L \phi_i'\phi_j' dx$$

$$+ p\sum_{j=0}^{N} \alpha_j \int_0^L \phi_i\phi_j' dx + q\sum_{j=0}^{N} \alpha_j \int_0^L \phi_i\phi_j dx = 0.$$ (B1.38)

Rearranging the terms of equation (B1.38), we have

$$p\alpha_0\phi_i(0) + \sum_{j=0}^{N} \alpha_j(-\int_0^L \phi_i'\phi_j' dx + p\int_0^L \phi_i\phi_j' dx + q\int_0^L \phi_i\phi_j dx) = pc_{in}\phi_i(0).$$ (B1.39)

Let

$$\phi_k = \begin{cases} (x - x_{k-1})/h, & x \in (x_{k-1}, x_k) \\ (x_{k+1} - x)/h, & x \in (x_k, x_{k+1}) \end{cases}, \quad \phi_k' = \begin{cases} 1/h, & x \in (x_{k-1}, x_k), \\ -1/h, & x \in (x_k, x_{k+1}) \end{cases}$$

for $k = 1, 2, \cdots, N-1$.

Then, equation (B1.38) forms a system of equations which leads to

$$Ac = b,$$ (B1.40)

where

$$a_{0,0} = a_{N,N} = -\frac{1}{h} + \frac{p}{2} + \frac{qh}{3},$$ (B1.41)

$$a_{i,i} = -\frac{2}{h} + \frac{2qh}{3}, \quad i = 1, 2, \cdots, N-1,$$ (B1.42)

$$a_{i,i-1} = \frac{1}{h} - \frac{p}{2} + \frac{qh}{6}, \quad i = 1, 2, \cdots, N-1,$$ (B1.43)

$$a_{i,i+1} = \frac{1}{h} + \frac{p}{2} + \frac{qh}{6}, \quad i = 0, 1, \cdots, N-2,$$ (B1.44)

$$b_0 = pc_{in},$$ (B1.45)

$$b_i = 0, \quad i = 1, 2, \cdots, N.$$ (B1.46)

For example, if $D=1$, $U=1$, $h=2.5$, $\gamma=0.2$, $c_{in}=100$ and $L=10$, the system becomes

$$\begin{bmatrix} -1.07 & -0.18 & & & \\ 0.82 & -1.13 & -0.18 & & \\ & 0.82 & -1.13 & -0.18 & \\ & & 0.82 & -1.13 & -0.18 \\ & & & 0.82 & -1.07 \end{bmatrix} \begin{bmatrix} c_0 \\ c_1 \\ c_3 \\ c_3 \\ c_4 \end{bmatrix} = \begin{bmatrix} -100 \\ 0 \\ 0 \\ 0 \\ 0 \end{bmatrix},$$

which can be solved by the Thomas algorithm to give

$$c_0 = 84.31,\ c_1 = 54.95,\ c_2 = 35.87,\ c_3 = 23.00,\ c_4 = 17.61.$$

These results (by FDM and FEM) are plotted in Figure B1.4. As expected, the concentration decreases due to the decay reaction as the chemical flows through the tank.

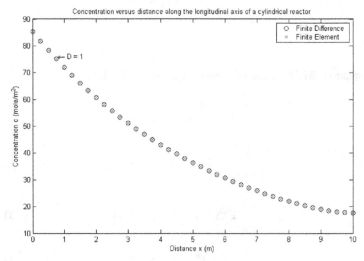

Figure B1.4. Concentration versus distance along the longitudinal axis of a cylindrical reactor for a chemical that decays with first-order decay kinetics with $h = 0.25$.

6. Interpretation and Conclusions

Clearly there is excellent agreement between the analytical and numerical results presented in Section 4 and the finite element results in Section 5 particularly, if we bear in mind, that only 4 elements have been used in the finite element approach in Section 5. It is also important to consider the case where dispersion is decreased. In this case, the curve would become steeper as mixing becomes less important relative to advection and decay. It should be noted if dispersion is decreased too much, the computation will become subject to numerical errors. This type of error is referred to as static instability. The criterion to avoid this static instability is

$$h \le \frac{2D}{U}.$$

Thus, it becomes more stringent (lower h) for cases where advection dominates over dispersion.

Figure B1.5 shows the results for static instability with $D = 0.1$, $U = 5$ and $h = 0.25$.

The mass fluxes for the steady state can be developed by substituting centered finite differences for the first derivative in the dispersion term, to give

$$J_i = -D\frac{c_{i+1} - c_{i-1}}{2h}. \tag{B1.47}$$

At the inlet node ($i = 0$),

$$J_0 = -D\frac{c_1 - c_{-1}}{2h}. \tag{B1.48}$$

c_{-1} can be removed by invoking the first boundary condition, which can be substituted to give

$$J_0 = Uc_{\text{in}} - Uc_0. \tag{B1.49}$$

A similar exercise can be performed for the outlet, where the original differential equation gives

$$J_n = -D\frac{c_{n+1} - c_{n-1}}{2h}. \tag{B1.50}$$

From the inspection of second boundary conditions we are led to conclude that $c_{n+1} = c_{n-1}$. In other words, mass fluxes $J_n = 0$ at the outlet. These results (analytical and numerical) are plotted in Figure B1.6 with $D = 1, 4$, $U = 1$ and $h = 0.25$.

7. Computer Algorithms

The computer algorithm used in Section 4.2 for the numerical solution simply involves finite difference discretization of the ODE together with solution of the resulting tridiagonal system of linear equations by the Thomas algorithm.

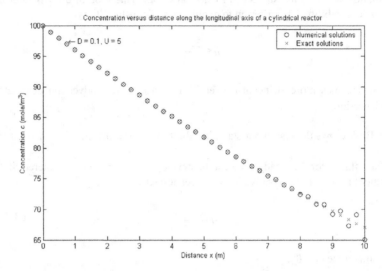

Figure B1.5. Concentration versus distance along the longitudinal axis of a cylindrical reactor for a chemical that decays with first-order decay kinetics subject to static instability.

A matrix A which has non-zero entries only on the diagonal, super-diagonal and the sub-diagonal is called a tridiagonal matrix. Such a matrix is commonly used in the numerical solution of differential equations. To solve such a system by traditional Gaussian elimination involves $O(n^3/3)$ operations. This is inefficient because the matrix itself is sparse, that is, contains many zero entries. In the following, we present an efficient algorithm to deal with such a system which involves only $O(n)$ operations.

Algorithm (*Thomas algorithm*): A tridiagonal system takes the form

$$Ax = \begin{pmatrix} \alpha_1 & \gamma_1 & 0 & \cdots & 0 \\ \beta_2 & \alpha_2 & \gamma_2 & \ddots & \vdots \\ 0 & \beta_3 & \alpha_3 & \ddots & 0 \\ \vdots & \ddots & \ddots & \ddots & \gamma_{n-1} \\ 0 & \cdots & 0 & \beta_n & \alpha_n \end{pmatrix} \begin{pmatrix} x_1 \\ x_2 \\ \vdots \\ x_{n-1} \\ x_n \end{pmatrix} = \begin{pmatrix} b_1 \\ b_2 \\ \vdots \\ b_{n-1} \\ b_n \end{pmatrix}.$$

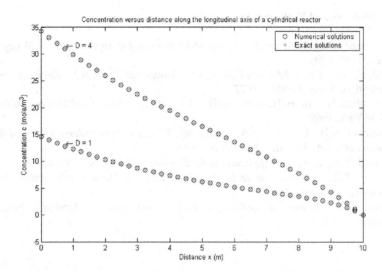

Figure B1.6. Mass fluxes versus distance along the longitudinal axis of a cylindrical reactor for a chemical that decays with first-order decay kinetics.

Applying LU decomposition of A, we have

$$A = \begin{pmatrix} \alpha_1 & \gamma_1 & 0 & \cdots & 0 \\ \beta_2 & \alpha_2 & \gamma_2 & \ddots & \vdots \\ 0 & \beta_3 & \alpha_3 & \ddots & 0 \\ \vdots & \ddots & \ddots & \ddots & \gamma_{n-1} \\ 0 & \cdots & 0 & \beta_n & \alpha_n \end{pmatrix} = \begin{pmatrix} l_1 & 0 & 0 & \cdots & 0 \\ \beta_2 & l_2 & 0 & \ddots & \vdots \\ 0 & \beta_3 & l_3 & \ddots & 0 \\ \vdots & \ddots & \ddots & \ddots & 0 \\ 0 & \cdots & 0 & \beta_n & l_n \end{pmatrix} \begin{pmatrix} 1 & \mu_1 & 0 & \cdots & 0 \\ 0 & 1 & \mu_2 & \ddots & \vdots \\ 0 & 0 & 1 & \ddots & 0 \\ \vdots & \ddots & \ddots & \ddots & \mu_{n-1} \\ 0 & \cdots & 0 & 0 & 1 \end{pmatrix}.$$

This is equivalent to

$$\alpha_1 = l_1,$$
$$\alpha_i = l_i + \beta_i \mu_{i-1}, \ i = 2, \ 3, \cdots, \ n,$$
$$l_i \mu_i = \gamma_i, \ i = 1, \ 2, \ \cdots, \ n-1.$$

The solution can be found efficiently by forward and back substitution

$$z_1 = b_1 / l_1, \ z_i = (b_i - \beta_i z_{i-1}) / l_i, \ i = 2, \ 3, \ \cdots, \ n,$$
$$x_n = z_n, \ x_i = (z_i - \mu_i z_{i+1}) / l_i, \ i = 1, \ 2, \ \cdots, \ n-1.$$

8. References and Bibliography

1. Allen, M.B. and Isaacson, E.L., *Numerical Analysis for Applied Science*, Wiley, New York, 1998.
2. Bajpai, A.C., Calus, I.M. and Fairley, J.A., *Numerical Methods for Engineers and Scientists*, Wiley, London, 1977.
3. Carnahan, B., Luther, H. A. and Wilkes J. O., *Applied Numerical Methods*, Wiley. New York, 1969.
4. Huebner, K.H., Thornton, E.A. and Byrom, T.G., *The Finite Element Method for Engineers*, 3rd ed., Wiley, New York, 1995.
5. Jewell, T.K., *Computer Applications for Engineers*, Wiley, New York, 1991.
6. Logan, D.L., *A First Course in the Finite Element Method*, PWS Publishing Company, Boston, 1997.
7. Rice, J. R., *Numerical Methods, Software, and Analysis*, Academic Press, Boston, 1983.

Project B2

THE FREE AND FORCED VIBRATION OF AN AUTOMOBILE

SUMMARY: This project considers the free and forced vibrations of an automobile supported by springs and shock absorbers. By using Newton's second law of motion, a mathematical model is formulated taking into account the damping force and the spring force. The case where the car is subject to periodic force is analyzed in detail. In this way it is possible to examine how the amplitude magnification factor varies with the ratio of the forcing and natural frequencies for a range of damping factors.

The realistic problem of determining the stability of a proposed design which has good comfort on rough roads in then considered. Of course, a number of modelling assumptions, including the use of realistic data, are required to achieve meaningful results. The numerical results are validated using MAPLE and MATLAB.

1. Background

Differential equations are often used to model the vibration of engineering systems. Some examples are a simple pendulum, a mass on a spring, and an inductance-capacitance electric circuit. The vibration of these systems may be damped by some energy-absorbing mechanism. In addition, the vibration may be free or subject to some external periodic disturbance. In the latter case the motion is said to be forced. In this project, we will examine the free and forced vibration of an automobile. The general approach is applicable to various other engineering problems.

2. Problem Statement

Examine the free and forced vibration of an automobile. Develop a mathematical model in the form of an ODE to represent the motion of the system.

Find a mathematical solution for systems which are (i) overdamped, (ii) underdamped, (iii) critically damped. Use x to represent the distance from the equilibrium position.

Now consider the case where the car is subject to a periodic force with forcing frequency ω given by $P = P_m \sin \omega t$ or $d = d_m \sin \omega t$, where $d_m = P_m / k =$ the static deflection of the car subject to a force P_m and k is the spring constant. Find an expression for the amplitude magnification factor denoted by x_m / d_m. Produce a graph to show a plot of magnification factor as a function of ω / p for various damping factors, where p represents the natural frequency of the undamped free vibration.

Determine the stability of a proposed design that has good comfort on rough roads. Assume the mass of the car is $m = 1.2 \times 10^6$ and it has a shock system with a damping coefficient of $c = 1 \times 10^7$. Further assume that the public's expectation of comfort is satisfied if the free vibration of the car is underdamped and the first crossing of the equilibrium position takes place in 0.05 sec. The stability of the car is considered satisfactory if at steady state the maximum distance $x = d_m$ is below 0.2m for all driving speeds.

Use MAPLE where possible to validate the numerical results.

3. Model Formulation

Figure B2.1. An automobile of mass m.

As shown in Figure B2.1, an automobile of mass m is supported by springs and shock absorbers. Shock absorbers offer resistance to the motion that is proportional to the vertical speed (up-and-down motion). Free vibrations result when the car is disturbed from equilibrium, such as after encountering a pothole. At any instant after hitting the pothole the net forces acting on the mass m are the resistance of the springs and the damping force of the shock absorbers. These forces tend to return the car to the original equilibrium state. According to Hooke's law, the resistance of the spring is proportional to the spring constant k and the distance from the equilibrium position, x. Therefore,

$$\text{Spring force} = -kx, \tag{B2.1}$$

where the negative sign indicates that the restoring force acts to return the car towards the position of equilibrium (that is, the negative x direction). The damping force of the shock absorbers is given by

$$\text{Damping force} = -c\frac{dx}{dt}, \tag{B2.2}$$

where c is a damping coefficient and dx/dt is the vertical velocity. The negative sign indicates that the damping force acts in the opposite direction against the velocity.

The equations of motion for the system are given by Newton's second law ($F = ma$), which for the present project is expressed as

$$m\frac{d^2x}{dt^2} = -c\frac{dx}{dt} + (-kx) \tag{B2.3}$$

$$\text{mass} \times \text{acceleration} = \text{damping force} + \text{spring force}$$

or

$$m\frac{d^2x}{dt^2} + c\frac{dx}{dt} + kx = 0. \tag{B2.4}$$

4. Mathematical/Numerical Solution

4.1 MATHEMATICAL SOLUTION

If we assume that the solution takes the form $x(t) = e^{rt}$, we can write the characteristic equation as

$$mr^2 + cr + k = 0. \tag{B2.5}$$

The unknown, r, is the solution of the quadratic characteristic equation that can be obtained either analytically or numerically. In this design problem, we will first use the analytical solution to give us general insight into the way the system motion is affected by the model coefficients m, c and k. We will also use numerical methods to obtain solutions and check the accuracy of the results using the analytical solution.

The solution of equation (B2.5) for r is given by the quadratic formula

$$\binom{r_1}{r_2} = \frac{-c \pm \sqrt{c^2 - 4mk}}{2m}.$$ (B2.6)

Note the significance of magnitude of c compared to $2\sqrt{km}$. If $c > 2\sqrt{km}$, r_1 and r_2 are negative real numbers, and the solution is of the form

$$x(t) = Ae^{r_1 t} + Be^{r_2 t},$$ (B2.7)

where A and B are constants to be determined from the initial conditions of x and dx/dt. Such systems are called *overdamped*.

If $c < 2\sqrt{km}$, the roots are complex,

$$\binom{r_1}{r_2} = \lambda \pm \mu i,$$

where

$$\mu = \frac{\sqrt{|c^2 - 4mk|}}{2m}$$

and the solution is of the form

$$x(t) = e^{-\lambda t}(A\cos \mu t + B\sin \mu t).$$ (B2.8)

Such systems are called *underdamped*.

Finally, if $c = 2\sqrt{km}$, the characteristic equation has a double root and the solution is of the form

$$x(t) = (A + Bt)e^{-\lambda t},$$ (B2.9)

where

$$\lambda = \frac{c}{2m}.$$

Such systems are called *critically damped*.

In all three cases, $x(t)$ approaches zero as t approaches infinity. This means that the car always returns to the equilibrium position after encountering the pothole.

The critical damping coefficient c_c is the value of c that makes the radical in equation (B2.6) equal to zero,

$$c_c = 2\sqrt{km} \text{ or } c_c = 2mp, \tag{B2.10}$$

where

$$p = \sqrt{\frac{k}{m}}. \tag{B2.11}$$

The ratio c/c_c is called the *damping factor* and p is called the *natural frequency* of the undamped free vibration.

Now, let us consider the case where the car is subject to a periodic force given by

$$P = P_m \sin \omega t \text{ or } d = d_m \sin \omega t \tag{B2.12}$$

where $d_m = P_m/k = $ the static deflection of the car subject to a force P_m. The governing differential equation for this case is

$$m\frac{d^2x}{dt^2} + c\frac{dx}{dt} + kx = P_m \sin \omega t. \tag{B2.13}$$

The general solution of this equation is obtained by adding a particular solution to the free vibration solution given by equations (B2.7)–(B2.9). Let us consider the steady-state motion of the forced system where the initial transient motion has been damped out. We assume that this steady-state solution has the form

$$x(t) = x_m \sin(\omega t - \phi). \tag{B2.14}$$

Hence

$$x(t) = x_m(\sin \omega t \cos \phi - \cos \omega t \sin \phi)$$

$$\frac{dx}{dt} = \omega x_m(\cos \omega t \cos \phi + \sin \omega t \sin \phi)$$

$$\frac{d^2x}{dt^2} = -\omega^2 x_m(\sin \omega t \cos \phi - \cos \omega t \sin \phi).$$

Substitution into equation (B2.13) leads to

$$-m\omega^2 x_m(\sin\omega t\cos\phi - \cos\omega t\sin\phi) + c\omega x_m(\cos\omega t\cos\phi + \sin\omega t\sin\phi)$$
$$+kx_m(\sin\omega t\cos\phi - \cos\omega t\sin\phi) = P_m\sin\omega t$$

and so

$$(k - \omega^2 m)x_m\cos\phi + c\omega x_m\sin\phi = P_m \tag{B2.15}$$
$$-(k - \omega^2 m)x_m\sin\phi + c\omega x_m\cos\phi = 0.$$

From the second equation of (B2.15) we have

$$\cos\phi = \frac{(k - \omega^2 m)\sin\phi}{c\omega}. \tag{B2.16}$$

Substituting this into the first equation of (B2.15) yields

$$\left[\frac{(k - \omega^2 m)^2}{c\omega} + c\omega\right]x_m\sin\phi = P_m$$

and hence

$$x_m\sin\phi = \frac{c\omega P_m}{[(k - \omega^2 m)^2 + c^2\omega^2]}.$$

This means that equation (B2.16) gives

$$x_m\cos\phi = \frac{(k - \omega^2 m)P_m}{(k - \omega^2 m)^2 + c^2\omega^2}.$$

Since

$$x_m^2\sin^2\phi + x_m^2\cos^2\phi = x_m^2,$$

we have

$$x_m^2 = \left[\frac{c\omega P_m}{(k - \omega^2 m)^2 + c^2\omega^2}\right]^2 + \left[\frac{(k - \omega^2 m)P_m}{(k - \omega^2 m)^2 + c^2\omega^2}\right]^2$$
$$= \frac{P_m^2}{(k - \omega^2 m)^2 + c^2\omega^2}$$

and hence

$$\frac{x_m}{P_m} = \frac{1}{\sqrt{(k - \omega^2 m)^2 + c^2 \omega^2}}.$$

Using equations (B2.10) and (B2.11) we obtain

$$\frac{x_m}{P_m} = \frac{1}{k\sqrt{(1-(\frac{\omega}{p})^2)^2 + 4(\frac{c}{c_c})^2(\frac{\omega}{p})^2}}$$

and hence

$$\frac{x_m}{d_m} = \frac{1}{\sqrt{(1-(\frac{\omega}{p})^2)^2 + 4(\frac{c}{c_c})^2(\frac{\omega}{p})^2}}. \tag{B2.17}$$

The quantity, x_m / d_m, called the *amplitude magnification factor*, depends only on the ratio of the actual damping to the critical damping and the ratio of the forcing frequency to the natural frequency. Note that when the forcing frequency ω approaches zero, the magnification factor approaches 1. Also if the system is lightly damped, that is, if c/c_c is small, then the magnification factor becomes large if ω is close to p. If the damping is zero, then the magnification factor becomes infinite when $\omega = p$, and the forcing function is said to be in resonance with the system. Finally, as ω/p becomes very large, the magnification factor approaches zero. Figure B2.2 shows a plot of magnification factor as a function of ω/p for various damping factors.

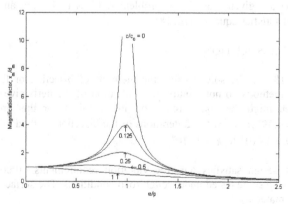

Figure B2.2. A plot of magnification factor as a function of ω/p for various damping factors.

Observe that the magnification factor may be kept small by choosing a large damping factor or by keeping the natural and the forced frequencies far apart.

The design of the car suspension system involves a trade-off between comfort and stability for all driving conditions and speeds. We are asked to determine the stability of a proposed design that has good comfort on rough roads. The mass of the car is $m = 1.2 \times 10^6$, and it has a shock system with a damping coefficient of $c = 1 \times 10^7$.

Assume that the public's expectation of comfort is satisfied if the free vibration of the car is underdamped and the first crossing of the equilibrium position takes place in $0.05s$. If, at $t = 0$, the car is suddenly displaced, x_0, from equilibrium and the velocity is zero $(dx/dt = 0)$, the solution of the equation of motion is given by equation (B2.8), with $A = x_0$ and $B = x_0 \lambda / \mu$. Therefore

$$x(t) = x_0 e^{-\lambda t} (\cos \mu t + \tfrac{\lambda}{\mu} \sin \mu t).$$

Our design conditions are met if

$$x(t) = 0 = \cos(0.05\mu) + \frac{\lambda}{\mu} \sin(0.05\mu)$$

or

$$0 = \cos\left(0.05 \sqrt{\frac{k}{m} - \frac{c^2}{4m^2}} \right) + \frac{c}{\sqrt{4km - c^2}} \sin\left(0.05 \sqrt{\frac{k}{m} - \frac{c^2}{4m^2}} \right). \qquad (B2.18)$$

Since c and m are given, our design problem reduces to finding an appropriate value of k that satisfies equation (B2.18).

4.2 NUMERICAL SOLUTION

Equation (B2.18) can be solved using the methods of bisection or false-position because these methods do not require the evaluation of the derivative of equation (B2.18), which might be considered a bit inconvenient for this problem. The solution is $k = 1.397 \times 10^9$, with 12 iterations of the bisection method with an initial bracket from $k = 1 \times 10^9$ to $k = 2 \times 10^9$.

Although this design satisfies our free vibration requirements (after hitting the pothole), it must also be tested under rough road conditions. The surface of the road can be approximated by

$$d = d_m \sin(2\pi x / D),$$

where d is the deflection, d_m is the maximum deflection of 0.1m, and D is the distance between peaks equal to 20m. If the horizontal speed of the car (m/s) is v, then the overall equation of motion for the system can be written as

$$m\frac{d^2x}{dt^2} + c\frac{dx}{dt} + kx = kd_m \sin(\frac{2\pi v}{D}t),$$

where $\omega = 2\pi v / D$ is the *forcing frequency*.

The stability of the car is considered satisfactory if at steady state the maximum distance $x = d_m$ is below 0.2m for all driving speeds. The damping factor is calculated according to equation (B2.10),

$$\frac{c}{c_c} = \frac{c}{2\sqrt{km}} = \frac{1\times10^7}{2\sqrt{(1.397\times10^9)(1.2\times10^6)}} = 0.1221 \cdot$$

Now, we seek values of ω / p that satisfy equation (B2.17), namely,

$$2 = \frac{1}{\left[\sqrt{\left(1-\frac{\omega^2}{p^2}\right)^2 + 4(0.1221)^2\left(\frac{\omega}{p}\right)^2}\right]} \cdot \qquad \text{(B2.19)}$$

When equation (B2.19) is expressed as a roots problem,

$$f\left(\frac{\omega}{p}\right) = 2\sqrt{\left(1-\frac{\omega^2}{p^2}\right)^2 + 4(0.1221)^2\left(\frac{\omega}{p}\right)^2} - 1 = 0, \qquad \text{(B2.20)}$$

we see that values of ω / p can be determined by finding the roots of equation (B2.20).

A plot of equation (B2.20) is shown in Figure B2.3. This plot shows that equation (B2.20) has two positive roots that can be determined by the bisection method using MATLAB. The smaller value for ω / p is found to equal 0.7300 in 18 iterations, with an estimated error of 0.000525% with lower and upper guesses of 0 and 1. The higher value of ω / p is found to be 1.1864 in 17 iterations, with an estimated error of 0.00064% with lower and upper guesses of 1 and 2.

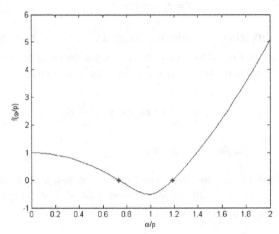

Figure B2.3. A plot of magnification factor as a function of ω/p for various damping factors.

It is also possible to express equation (B2.19) as a polynomial equation

$$\left(\frac{\omega}{p}\right)^4 - 1.9404\left(\frac{\omega}{p}\right)^2 + 0.75 = 0 \cdot \tag{B2.21}$$

This validates the result obtained using bisection. It also suggests that, although it superficially appears to be a fourth-order equation in ω/p, equation (B2.21) is actually a quadratic equation in $(\omega/p)^2$.

The value of the natural frequency p is given by equation (B2.11), namely,

$$p = \sqrt{\frac{1.397 \times 10^9}{1.2 \times 10^6}} = 34.12 \ s^{-1}.$$

The forcing frequencies, for which the maximum deflection is 0.2m, are then calculated as

$$\omega = 0.7300(34.12) = 24.91 \ s^{-1}$$
$$\omega = 1.1864(34.12) = 40.48 \ s^{-1},$$

which yield, using the equation

$$v = \frac{\omega D}{2\pi}$$

velocities of 285km/h and 464km/hr, respectively.

5. Model Validation

Both MATLAB and MAPLE been used to validate the numerical results in Section 4.2. For example, the values of ω/p from equation (B2.20) are obtained by MAPLE in Figure B2.4.

```
> m:=1.2*10^6;
  c:=1*10^7;
  k:=fsolve(cos(0.05*sqrt(k/m-c^2/(4*m^2)))+c*sin(0.05*sqrt(k/m-
  c^2/(4*m^2)))/sqrt(4*k*m-c^2),k);
  c_c:=2*sqrt(k*m);
  p:=sqrt(k/m);
  solve({2*sqrt((1-x^2)^2+4*(c/c_c)^2*x^2)-1,x>0});
```
$$m := 0.12000000\ 10^7$$
$$c := 10000000$$
$$k := 0.1396991571\ 10^{10}$$
$$c\_c := 0.8188748098\ 10^8$$
$$p := 34.11978373$$
$$\{x = 0.7299772551\}, \{x = 1.186373134\}$$

Figure B2.4. MAPLE validation of values of ω/p from equation (B2.20).

This project has examined the free and forced vibration of an automobile. It has dealt with the cases which are (i) overdamped, (ii) underdamped and (iii) critically damped. The case where the car is subject to a periodic force is investigated in detail. Consequently consideration has been given to the stability of a proposed design which has good comfort on rough roads.

Thus, using the above results and Figure B2.3, it is found that the proposed car design will behave acceptably for common driving speeds. At this point, the designer must be aware that the design would not meet suitability requirements at extremely high speeds (e.g., racing).

This design problem has allowed us to obtain some analytical results that were used to validate the accuracy of our numerical methods for finding roots. Real cases are normally more complicated which means that solutions can be obtained only by using numerical methods.

6. Interpretation and Conclusions

This project shows the importance of differential equations in the modelling of the vibration of an engineering system. In this case the system is an automobile supported by springs and shock absorbers. The shock absorbers offer resistance to the motion and free vibrations result when the car is disturbed from equilibrium, e.g., encountering a pothole in the road. It has been possible to devise a mathematical model for the system using both Hooke's law and Newton's second law of motion. This project includes a detailed analysis for the case when the car is subject to a periodic force. As a result graphs can be plotted to show how the amplitude magnification factor depends on the ratio of the actual damping to the critical damping and the ratio of the forcing frequency to the natural frequency. A realistic case has been considered which looked at the stability of a proposed design which leads to good comfort on rough roads.

This work is relevant to the vibration of other types of engineering systems. Examples include the simple pendulum, a mass-spring system and electric circuits involving inductance and capacitance. The vibration of all of these systems may be damped by some energy absorbing mechanism. Also, the vibration may be free or subject to some external periodic disturbance. In the latter case, the motion is said to be forced.

7. Computer Algorithms

The computer algebraic systems MAPLE and MATLAB were sufficient to handle the numerical aspects of this project.

8. References and Bibliography

1. Hayt, W.H. and Kemmerly, J.E., *Engineering Circuit Analysis*, McGraw-Hill, New York, 1986.
2. Jaluria, Y., *Computer Methods for Engineering*, Allyn and Bacon, Inc., Boston, 1988.
3. James, G., *Modern Engineering Mathematics*, 2nd ed., Addison-Wesley, London, 1996.
4. Köckler, N., *Numerical Methods and Scientific Computing*, Oxford University Press, New York, 1994.
5. Kreyszig, E., *Advanced Engineering Mathematics*, 7th ed., Wiley, New York, 1993.
6. Yakowitz, S. and Szidarovsky, F., *An Introduction to Numerical Computation*, Macmillan, New York, 1986.

Project B3

CANTILEVER BEAM SUBJECTED TO AN END LOAD

SUMMARY: This project investigates the mathematical modelling of a beam which is either imbedded at both ends or free at one end. The model involves a fourth-order ODE together with boundary conditions which depend on the manner in which the beam is supported. Analytical solutions are obtained for a number of test cases.

The particular case of a steel cantilever beam subjected to an end load is then investigated by using both analytical and numerical techniques. The deflection of the beam obtained by an analytical approach is validated by using finite difference methods and suggestions are given on possible finite element approaches.

1. Background

Consider a horizontal beam AB as shown in Figure B3.1 with the assumption that the beam is uniform in cross section and of homogeneous material. If there is no load, the axis of symmetry is the straight line which is indicated by the solid line.

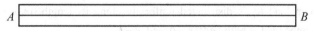

A B

Figure B3.1. A horizontal beam.

However, when there are external loadings, the beam is distorted and the result is a curve called the deflection curve or elastic curve. Beams can be supported in many ways. A cantilever beam has one end rigidly fixed while the other end is free to move as shown in Figure B3.2(a). A beam which is supported at both ends A and B is called a simply supported beam (see Figure B3.2 (b)).

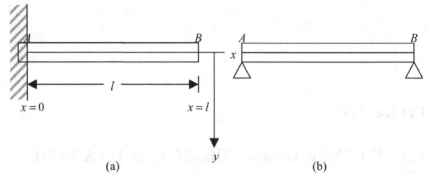

(a)   (b)
Figure B3.2(a). A cantilever beam. *(b)* A simply supported beam.

Using calculus and the Bernoulli-Euler law, it is possible to develop a governing differential equation to represent the deflection of the beam. Also, it is possible to formulate boundary conditions associated with this differential equation which depend on the manner in which the beam is supported.

2. Problem Statement

Consider first the general problem of deflection of beams. Show that the deflection $y(x)$ of a beam of length l subjected to a vertical load $W(x)$ such that x denotes the distance from one end, satisfies the differential equation:

$$EI\frac{d^4y}{dx^4} = W(x), \ 0 < x < l, \tag{B3.1}$$

where the modulus of elasticity (E) and moment of inertia (I) are known constants for a particular beam.

Discuss the possible boundary conditions for problems of this type.

Determine the deflection $y(x)$ of the beam in the following cases:

(a) A constant load W_0 is distributed uniformly along its length $0 \le x \le l$ and the beam is imbedded at both ends $x = 0$ and $x = l$.

(b)
$$W(x) = \begin{cases} W_0\left(1 - \frac{2}{l}x\right) & , \ 0 < x < \frac{l}{2}, \\ 0 & , \ \frac{l}{2} < x < l, \end{cases} \tag{B3.2}$$

and the beam is imbedded at both ends $x = 0$ and $x = l$.

(c) A constant load W_0 is distributed uniformly along its length $0 \le x \le l$ and the beam is imbedded at $x = 0$ and free at $x = l$.

(d) A load W_0 is concentrated at $x = l$ and the beam is imbedded at $x = 0$ and free at $x = l$.

Now consider the particular case of a steel cantilever beam subjected to an end load of 35,000N (see Figure B3.3 later). Assume the following data:

Modulus of elasticity, E	$= 0.210 \text{N/m}^2$
Breadth of beam, b	$= 0.04 \text{m}$
Depth of beam, d	$= 0.20 \text{m}$
Overall length of beam, l	$= 1.00 \text{m}$
Load on the beam, W_0	$= 35,000 \text{N}$
Poisson's ratio, v	$= 0.25$

Formulate a mathematical model stating clearly any assumptions made. Hence find the maximum deflection, y_{max}, of the cantilever beam together with the bending moment, M, and the maximum stress, σ_{max}. Comment on the accuracy of this value of y_{max} and suggest how it can be improved. Validate the model by using a finite difference approximation and suggest possible finite element approaches.

3. Model Formulation

To formulate a mathematical model and develop the governing differential equation, we consider here a cantilever beam which is built-in at the end $x = 0$, and the end $x = l$ is free to move. The deflection y is measured positive downward as shown in Figure B3.2 (a).

The Bernoulli-Euler law states that the curvature ρ of the beam is proportional to the bending moment M, i.e.,

$$\rho \alpha M$$

or

$$\rho = \frac{M}{EI}, \tag{B3.3}$$

where E is the Young's modulus of elasticity and depends on the material used in designing the beam and I is the moment of inertia of the cross-section of the beam at x with respect to a horizontal line passing through the centre of gravity of this cross-section. The quantity EI is a measure of the flexural rigidity of the beam.

From the calculus, the curvature is given by

$$\rho = \frac{1}{\left\{1+\left(\dfrac{dy}{dx}\right)^2\right\}} \cdot \frac{d^2y}{dx^2}. \qquad (B3.4)$$

If the deflection y is small, then so is the slope $\dfrac{dy}{dx}$ and therefore the curvature is approximately equal to $\dfrac{d^2y}{dx^2}$.

Thus, from equations (B3.3) and (B3.4), we have

$$\frac{d^2y}{dx^2} = \frac{M}{EI}. \qquad (B3.5)$$

But, the shear force S is given by

$$S = \frac{dM}{dx} \qquad (B3.6)$$

and the load per unit length of the beam, $W(x)$, is related to the shear force by

$$W(x) = \frac{dS}{dx}. \qquad (B3.7)$$

If there is no other load on the beam, then

$$W(x) = \frac{dS}{dx} = \frac{d^2M}{dx^2} = \frac{d^2}{dx^2}\left(EI\frac{d^2y}{dx^2}\right), \qquad (B3.8)$$

which is subsequently written as

$$\frac{d^4y}{dx^4} = \frac{W(x)}{EI}, \quad 0<x<l, \qquad (B3.9)$$

where E and I are given constants.

This is a fourth-order differential equation, the general solution of which will contain four arbitrary constants. Therefore, four boundary conditions must be given.

For the problem considered here, the following conditions would be suitable:

At $x = 0$: $$y(0) = 0 \tag{B3.10}$$

$$y'(0) = 0 \tag{B3.11}$$

At $x = l$: $$y''(l) = 0 \tag{B3.12}$$

$$y'''(l) = 0, \tag{B3.13}$$

where y', y'' and y''' denote the first, second and third derivatives of y with respect to x.

Conditions (B3.10) and (B3.11) state that there is no deflection at the built-in or imbedded end, $x = 0$, and the slope of the tangent to the elastic curve at $x = 0$ is zero, respectively. Also, conditions (B3.12) and (B3.13) are obtained from the fact that the curvature of the elastic curve is zero at the free end, $x = l$, and the shearing force vanishes at the free end, $x = l$, respectively.

In general, the boundary conditions associated with the differential equation (B3.9) depend on the manner in which the beam is supported. The most common boundary conditions are as follows:

(1) Fixed end, built-in, imbedded, clamped end

$$y = y' = 0.$$

(2) Hinged or simply supported end

$$y = y'' = 0.$$

(3) Free end

$$y'' = y''' = 0.$$

This problem is an example of a two point boundary value problem in contrast to the initial value problem in which conditions are specified at a single point.

4. Mathematical Solution

4.1 TEST EXAMPLE (A)

In this case a constant load W_0 is distributed uniformly along the length of the beam, $0 \le x \le l$, and the beam is imbedded at both ends, $x = 0$ and $x = l$.

In this case, the deflection $y(x)$ must satisfy the boundary conditions:

$$y(0) = 0, \ y(l) = 0, \ y'(0) = 0, \ y'(l) = 0. \tag{B3.14}$$

The first two conditions indicate that there is no vertical deflection at the ends; the last two conditions mean that the line of the deflection is horizontal (zero slope) at the ends.

Since a constant load W_0 is uniformly distributed along the beam, we have

$$W(x) = W_0, \ 0 < x < l. \tag{B3.15}$$

One possible approach is to use Laplace transforms. Transforming the equation (B3.9) gives

$$s^4 Y(s) - s^3 y(0) - s^2 y'(0) - sy''(0) - y'''(0) = \frac{W_0}{EIs}, \tag{B3.16}$$

where $Y(s) \equiv \mathcal{L}\{y(x)\}$.

Taking $A = y''(0)$ and $B = y'''(0)$, then

$$Y(s) \equiv \frac{A}{s^3} + \frac{B}{s^4} + \frac{W_0}{EIs^5}$$

since $y(0) = y'(0) = 0$ from equation (B3.14).

Inverting gives

$$y(x) = \frac{A}{2!} \mathcal{L}^{-1}\left\{\frac{2!}{s^3}\right\} + \frac{B}{3!} \mathcal{L}^{-1}\left\{\frac{3!}{s^4}\right\} + \frac{W_0}{4!EI} \mathcal{L}^{-1}\left\{\frac{4!}{s^5}\right\}$$

$$= \frac{Ax^2}{2} + \frac{Bx^3}{6} + \frac{W_0 x^4}{24EI},$$

where $\mathcal{L}^{-1}$ denotes the inverse Laplace transform.

Applying the given conditions $y(l) = y'(l) = 0$ gives

$$\frac{Al^2}{2} + \frac{Bl^3}{6} + \frac{W_0 l^4}{24EI} = 0$$

$$Al + \frac{Bl^2}{2} + \frac{W_0 l^3}{6EI} = 0.$$

Solving gives $A = \dfrac{W_0 l^2}{12EI}$ and $B = -\dfrac{W_0 l}{2EI}$.

Hence

$$y(x) = \frac{W_0 l^2 x^2}{24EI} - \frac{W_0 l x^3}{12EI} + \frac{W_0 x^4}{24EI}$$

$$= \frac{W_0 x^2 (x-l)^2}{24EI}. \qquad (B3.16)$$

4.2 TEST EXAMPLE (B)

In this case the distribution of the load, $W(x)$, along the beam is given by equation (B3.2) and the boundary conditions are as follows:

$$y(0) = 0, \ y'(0) = 0, \ y(l) = 0, \ y'(l) = 0. \qquad (B3.17)$$

Using the unit step function $U(x)$, where

$$U(x) = \begin{cases} 0 & , x < 0 \\ 1 & , x > 0, \end{cases} \qquad (B3.18)$$

we can write

$$W(x) = W_0\left(1 - \frac{2x}{l}\right) - W_0\left(1 - \frac{2x}{l}\right)U\left(x - \frac{l}{2}\right)$$

from equation (B3.2). Hence

$$W(x) = \frac{2W_0}{l}\left\{\frac{l}{2} - x + \left(x - \frac{l}{2}\right)U\left(x - \frac{l}{2}\right)\right\}.$$

So our governing differential equation becomes

$$\frac{d^4 y}{dx^4} = \frac{2W_0}{EII}\left\{\frac{l}{2} - x + \left(x - \frac{l}{2}\right)U\left(x - \frac{l}{2}\right)\right\}. \tag{B3.19}$$

Taking Laplace transforms gives

$$s^4 Y(s) - s^3 y(0) - s^2 y'(0) - s y''(0) - y'''(0)$$
$$= \frac{2W_0}{EII}\left\{\frac{l}{2s} - \frac{1}{s^2} + \frac{1}{s^2}e^{-\frac{ls}{2}}\right\},$$

where $\mathcal{L}\{y(x)\} \equiv Y(s)$.

Taking $A = y''(0)$ and $B \equiv y'''(0)$, then

$$Y(s) \equiv \frac{A}{s^3} + \frac{B}{s^4} + \frac{2W_0}{EII}\left\{\frac{l}{2s^5} - \frac{1}{s^6} + \frac{1}{s^6}e^{-\frac{ls}{2}}\right\}$$

since $y(0) = y'(0) = 0$ from equation (B3.17).

Inverting gives

$$y(x) = \frac{A}{2!}\mathcal{L}^{-1}\left\{\frac{2!}{s^3}\right\} + \frac{B}{3!}\mathcal{L}^{-1}\left\{\frac{3!}{s^4}\right\}$$
$$+ \frac{2W_0}{EII}\left[\frac{l}{2 \times 4!}\mathcal{L}^{-1}\left\{\frac{4!}{s^5}\right\} - \frac{1}{5!}\mathcal{L}^{-1}\left\{\frac{5!}{s^6}\right\} + \frac{1}{5!}\mathcal{L}^{-1}\left\{\frac{5!}{s^6}e^{-\frac{ls}{2}}\right\}\right]$$
$$= \frac{Ax^2}{2} + \frac{Bx^3}{6} + \frac{W_0}{60EII}\left[\frac{5lx^4}{2} - x^5 + \left(x - \frac{l}{2}\right)^5 U\left(x - \frac{l}{2}\right)\right],$$

where $\mathcal{L}^{-1}$ denotes the inverse Laplace transform.

Applying the given conditions $y(l) = y'(l) = 0$ gives

$$\frac{Al^2}{2} + \frac{Bl^3}{6} + \frac{49W_0 l^4}{1920EI} = 0$$
$$Al + \frac{Bl^2}{2} + \frac{85W_0 l^3}{960EI} = 0.$$

Solving gives $A = \dfrac{23W_0 l^2}{960EI}$ and $B = -\dfrac{9W_0 l}{40EI}$.

Hence

$$y(x) = \frac{23W_0 l^2 x^2}{1920EI} - \frac{3W_0 l x^3}{80EI}$$

$$+ \frac{W_0}{60EI l}\left[\frac{5lx^4}{2} - x^5 + \left(x - \frac{l}{2}\right)^5 U\left(x - \frac{l}{2}\right)\right]. \tag{B3.20}$$

4.3 TEST EXAMPLE (C)

In this case a constant load W_0 is distributed uniformly along the length of the beam, $0 \le x \le l$, and the beam is imbedded at $x = 0$ and free at $x = l$.

In this case the deflection $y(x)$ must satisfy the boundary conditions:

$$y(0) = 0,\ y'(0) = 0,\ y''(l) = 0,\ y'''(l) = 0. \tag{B3.21}$$

Again, we could use Laplace transforms, but in this case it is possibly easier to simply integrate the governing differential equation

$$\frac{d^4 y}{dx^4} = \frac{W_0}{EI} \tag{B3.22}$$

and impose the boundary conditions given in equation (B3.21). Integration of equation (B3.22) gives

$$y''' = \frac{W_0 x}{EI} + A.$$

Using condition $y'''(l) = 0$ gives

$$A = -\frac{W_0 l}{EI}.$$

Integrating again gives

$$y'' = \frac{W_0}{2EI} x^2 + Ax + B.$$

Using condition $y''(l) = 0$ gives

$$B = -\frac{W_0 l^2}{2EI} + \frac{W_0 l^2}{EI} = \frac{W_0 l^2}{2EI}.$$

Integrating again gives

$$y' = \frac{W_0 x^3}{6EI} + \frac{Ax^2}{2} + Bx + C.$$

Using condition $y'(0)$ yields

$$C = 0$$

and the final integration gives

$$y = \frac{W_0 x^4}{24EI} + \frac{Ax^3}{6} + \frac{Bx^2}{2} + D.$$

The condition $y(0) = 0$ yields

$$D = 0.$$

Thus the solution can be written as

$$\begin{aligned} y(x) &= \frac{W_0 x^4}{24EI} - \frac{W_0 lx^3}{6EI} + \frac{W_0 l^2 x^2}{4EI} \\ &= \frac{W_0}{24EI}(x^4 - 4lx^3 + 6l^2 x^2). \end{aligned} \quad (B3.23)$$

Hence the deflection at the free end, $x = l$, is given by

$$y = \frac{W_0}{24EI}(l^4 - 4l^4 + 6l^4) = \frac{W_0 l^4}{8EI}. \quad (B3.24)$$

4.4 TEST EXAMPLE (D)

In this case a load W_0 is concentrated at $x = l$ which means that we must introduce the unit impulse or Dirac delta function $\delta(x-l)$, defined by

$$\delta(x-l) = \lim_{\varepsilon \to 0} \left[\frac{1}{\varepsilon} \{ U(x) - U(x-l) \} \right]. \tag{B3.25}$$

This means that $\delta(x-l) = 0$ for all $x \neq l$ and that

$$\int_{-\infty}^{\infty} \delta(x-l)dx = 1.$$

Hence, the governing differential equation in this case is

$$\frac{d^4 y}{dx^4} = \frac{W_0}{EI} \delta(x-l) \tag{B3.26}$$

and the boundary conditions are as follows:

$$y(0) = 0, \ y'(0) = 0, \ y''(l) = 0, \ y'''(l) = 0. \tag{B3.27}$$

Again, applying Laplace transforms gives

$$s^4 Y(s) - s^3 y(0) - s^2 y'(0) - sy''(0) - y'''(0) = \frac{W_0}{EI} e^{-sl},$$

where $Y(s) \equiv \mathcal{L}\{y(x)\}$.

Taking $A = y''(0)$ and $B \equiv y'''(0)$, then

$$Y(s) \equiv \frac{A}{s^3} + \frac{B}{s^4} + \frac{W_0}{EIs^4} e^{-sl}$$

since $y(0) = y'(0) = 0$ from equation (B3.27).

Inverting gives

$$y(x) = \frac{A}{2!} \mathcal{L}^{-1} \left\{ \frac{2!}{s^3} \right\} + \frac{B}{3!} \mathcal{L}^{-1} \left\{ \frac{3!}{s^4} \right\} + \frac{W_0}{3!EI} \mathcal{L}^{-1} \left\{ \frac{3!e^{-sl}}{s^4} \right\}$$

$$= \frac{Ax^2}{2} + \frac{Bx^3}{6} + \frac{W_0(x-l)^3}{6EI},$$

where $\mathcal{L}^{-1}$ denotes the inverse Laplace transform.

Applying the given conditions $y''(l) = y'''(l) = 0$ gives

$$A + Bl = 0$$

$$B + \frac{W_0}{EI} = 0.$$

Solving gives $A = \frac{W_0 l}{EI}$ and $B = -\frac{W_0}{EI}$.

Hence

$$y(x) = \frac{W_0 l x^2}{2EI} - \frac{W_0 x^3}{6EI} + \frac{W_0 (x-l)^3}{6EIl} = \frac{W_0 x^2 (3l-x)}{6EI}. \qquad \text{(B3.28)}$$

The deflection at the free end, $x = l$, is given by

$$y = \frac{W_0 l^3}{3EI}. \qquad \text{(B3.29)}$$

4.5 STEEL CANTILEVER BEAM SUBJECTED TO AN END LOAD

Now we consider the realistic case of a steel cantilever beam subjected to an end load of 35,000N (see Figure B3.3). In Test Example (D) in Section 4.4 we have found analytically an expression for the deflection $y(x)$ of the beam at a distance x along the beam, namely

$$y(x) = \frac{W_0 x^2}{6EI} (3l - x), \qquad \text{(B3.30)}$$

with the maximum deflection occurring at the end at which the load is applied and given by

$$y_{max} = \frac{Wl^3}{3EI}. \qquad \text{(B3.31)}$$

In this realistic problem, we use the following data:

Modulus of elasticity, E $= 210,000\text{N/mm}^2$
Load on the beam, W_0 $= 35,000\text{N}$
Overall length of beam, l $= 1,000\text{mm}.$

The moment of inertia, I, of the cross-section about the neutral axis is given by

$$I = \frac{bd^2}{12},$$
(B3.32)

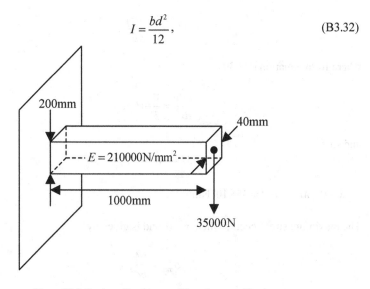

Figure B3.3. Steel cantilever beam subjected to an end load.

where

$$b = \text{breadth of beam (40mm)}$$
$$d = \text{depth of beam (200mm)}.$$

Hence, the moment of inertia is given by

$$I = \frac{(40)(200)^3}{12} \text{mm}^4$$

and so the maximum deflection is

$$y_{max} = \frac{35,000 \times (1000)^3 \times 12}{3 \times 21 \times 10^4 \times 40 \times (200)^3}$$
$$= 2.083\text{mm}.$$

In the above analysis it is assumed that the applied load produces a distortion in pure bending only. However, shear stresses act across the cross-section and these produce an additional deflection of the centroid of the end section. This small additional deflection due to shear has been ignored in the above calculation.

The bending moment M is given by

$$M = -EI\frac{d^2y}{dx^2},$$ (B3.33)

where from equation (B3.30)

$$\frac{d^2y}{dx^2} = \frac{W_0}{EI}(l-x)$$

and so

$$M = -W_0(l-x).$$ (B3.34)

At $x = 0$, $M = -W_0 l = 35 \times 10^6 \, \text{Nmm}$.

The maximum stress occurs when $x = 0$ and is given by

$$\sigma_{max} = \frac{6M}{bd^2}$$

as illustrated in Figure B3.4. Thus $\sigma_{max} = 131.25 \text{N/mm}^2$.

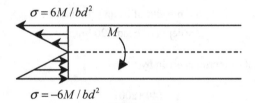

Figure B3.4. The maximum stress at $x = 0$

5. Model Validation

5.1 FINITE DIFFERENCE METHOD

A finite difference approximation is used to solve the bending moment equation

$$EI\frac{d^2y}{dx^2} = W_0(l-x)$$ (B3.35)

subject to the boundary conditions

$$y(0) = 0, \ y'(0) = 0. \tag{B3.36}$$

The beam is divided into n equal parts of length h as shown in Figure B3.5. The deflection at the ends of the parts are $y_0, y_1, y_2, \cdots, y_n$. The second derivative in equation (B3.35) is approximated at x_i by the expression

$$\frac{y_{i+1} - 2y_i + y_{i-1}}{h^2}$$

leading to the approximate bending moment equation

$$y_{i+1} - 2y_i + y_{i-1} = \frac{W_0 h^2}{EI}(l - x_i). \tag{B3.37}$$

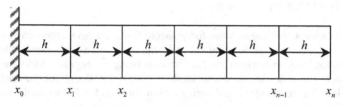

Figure B3.5. Division of the beam into n equal parts.

Remembering that $y_0 = 0$ and adding a fictitious deflection $y_{-1} = y_1$ to give $y'(0) = 0$, we obtain the following system of linear equations

$$
\begin{aligned}
y_0 &= 0 \\
2y_1 &= C(l - x_0) \\
y_2 - 2y_1 &= C(l - x_1) \\
y_3 - 2y_2 + y_1 &= C(l - x_2) \\
&\vdots \\
y_n - 2y_{n-1} + y_{n-2} &= C(l - x_{n-1}),
\end{aligned} \tag{B3.38}
$$

where $C = \dfrac{W_0 h^2}{EI}$.

These equations are solved for $n = 4, 8, 16, 32$ and 64 and the results are compared with the exact value given by equation (B3.30) up to 5 decimal places. Selected values of x are given in Table B3.1.

Table B3.1. Finite difference results.

x	0	250	500	750	1000
$y(n=4)$	0	0.19531	0.68359	1.36719	2.14844
$y(n=8)$	0	0.18311	0.65918	1.33057	2.09961
$y(n=16)$	0	0.18005	0.65308	1.32141	2.08740
$y(n=32)$	0	0.17929	0.65155	1.31912	2.08435
$y(n=64)$	0	0.17910	0.65117	1.31855	2.08359
y(exact)	0	0.17904	0.65104	1.31836	2.08833

To obtain results close to the exact values, we require to use $n = 64$. Of course, the finite difference approximation used is only $O(h^2)$ and more accurate results could be obtained by using higher order finite difference approximations.

5.2 FINITE ELEMENT METHOD

Below we give some suggestions for possible finite element approaches without going to the extent of performing the calculations. Figure B3.6 shows the beam approximated by 4 finite elements, the elements being rectangular blocks each with eight nodes. The nodes are numbered as shown in Figure B3.6. Note that each finite element is connected to the neighbouring one by three nodes. For example, nodes 11, 12 and 13 connect the second and third finite elements of the beam.

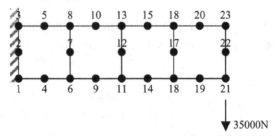

Figure B3.6. The beam approximated by 4 rectangular finite elements.

The second possibility would be to use the simple element commonly used in two dimensional plane stress problems, namely, the rectangular four noded element. The structure can then be divided into rectangular elements with a typical element as shown in Figure B3.7. The element with nodes i, j, k, l is assumed to have nodal coordinates (x_i, y_i), (x_j, y_j), (x_k, y_k), (x_l, y_l). Then a displacement function is chosen to provide the displacements within the element, given the displacements at the nodes: $(u_i, v_i, u_j, v_j, u_k, v_k, u_l, v_l)$,

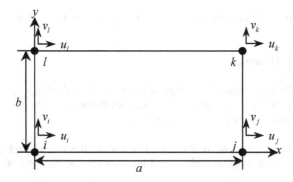

Figure B3.7. Structure of the rectangular four noded element.

In this case it would be convenient to take the element as having sides parallel to the x and y axes of length a and b as shown in Figure B3.7 and choose the displacement function given by polynomials of the form

$$u = c_1 + c_2 x + c_3 y + c_4 xy$$
$$v = c_5 + c_6 x + c_7 y + c_8 xy. \tag{B3.39}$$

The coefficients in equation (B3.39) can be obtained by substituting in turn the nodal coordinates and solving resulting systems of linear equations.

Of course, in the application of the finite element method it is not necessary to actually perform the calculations but merely to input the nodal coordinates and the type of element we wish to use. However, it is necessary to be aware of the displacement functions that are being assumed within the chosen finite element computer package and the consequences involved.

6. Interpretation and Conclusions

This project has looked at the general problem of deflection of beams. A mathematical model has been formulated and a number of test cases have been considered, including (i) constant load distributed uniformly along the length of the beam; (ii) prescribed load which varies with the distance along the beam; (iii) concentrated load at a single point on the beam.

Analytical solutions have been obtained for cases where the beam is imbedded at both ends or is free at one end.

The realistic case of a steel cantilever beam (with prescribed parameters) subjected to an end load has been solved numerically. The results obtained from finite difference methods agree well with the analytical solution. Suggestions are also included to indicate how the finite element method can be used to deal with the above problem.

7. Computer Algorithms

The finite difference method described in Section 5.1 resulted in a system of linear equations (B3.38). A computer program was developed to solve this system using forward substitution for a range of values of n.

There are quite a number of available finite element "black box" packages to solve the cantilever beam problem with an end load, which use the suggestions given in Section 5.2.

8. References and Bibliography

1. Bajpai, A.C., Mustoe, L.R. and Walker, D., *Advanced Engineering Mathematics*, Wiley, New York, 1978.
2. Barber, J.R., *Elasticity*, Kluwer Academic Publishers, Dordrecht, 1992.
3. Burden, R.L. and Faires, J.D., *Numerical Analysis*, 5th ed., PWS Publishing Company, Boston, 1993.
4. Jaluria, Y., *Computer Methods for Engineering*, Allyn and Bacon, Inc., Boston, 1988.
5. Köckler, N., *Numerical Methods and Scientific Computing*, Oxford University Press, New York, 1994.
6. Koenig, H.A., *Modern Computational Methods*, Taylor & Francis, London, 1998.
7. Timoshenko, S.P. and Goodier, J.N., *Theory of Elasticity*, 3rd ed., McGraw-Hill, New York, 1987.

ODE Problems

1. The rate of change of the concentration of pollution in a lake is equal to the difference between the concentration of polluted water entering the lake and that leaving the lake. Assume that water containing a constant concentration of C kg/km^3 of pollutants enters the lake at a rate of $150\,km^3/year$, and water leaves the lake at the same rate. Also assume that the volume of the lake remains constant at $5000\,km^3$.

   (a) Formulate a mathematical model to represent the rate of change of concentration of pollution in the lake. Find a mathematical solution.

   (b) If the initial concentration of pollution is $40\,kg/km^3$, find the particular solution to the problem.

   (c) The fastest possible cleanup of the lake will occur if all pollution inflow ceases. This is represented by $C = 0$. If all pollution into the lake was stopped immediately, how long would it take to reduce pollution to 50% of its current value?

   (d) Use the computer to graph your solution for the first 100 years after pollution stops. What happens to the concentration as time goes on?

2. A projectile of mass 0.20kg is shot vertically upward with an initial velocity of 10 m/sec. It is then slowed down due to the forces exerted by gravity and air resistance.

(a) If the force due to air resistance equals 0.005 times the square of the projectile's instantaneous velocity acting in the opposite direction to the velocity, produce a mathematical model using an initial-value differential equation. Use velocity as the dependent variable. Solve to find an expression for velocity in terms of time t.

(b) Apply a fourth-order Runge-Kutta method with $h = 0.05$ to estimate the projectile's instantaneous velocity for time $t = 0.05(0.05)1.00$ sec. Validate your results using the exact solution from (a).

3. An automobile shock absorber coil spring system is designed to support 800lb, the portion of the automobile's weight it supports. The spring has a constant of 50 slugs/in. The effect of a bumpy road on the system can be described by the periodic function

$$f(t) = 300 \sin 4t \text{ (in slug} - \text{in} / \sec^2),$$

which acts upward on the tyre. The system is initially in equilibrium at rest.

(a) Assume that the automobile's shock absorber is so worn that it provides no effective damping force. Find a particular solution which describes the vertical displacement of the automobile over time. Use the computer to graph the particular solution for the first ten seconds of motion. Describe the system's performance.

(b) Now assume that the shock absorber is replaced. The new shock absorber exerts a damping force (in pounds) which is equal to 50 times the instantaneous vertical velocity of the system (in inches per second). Model this improved system with an initial value problem. Solve it subject to the conditions described in part (a). Use the computer to graph the resulting equation for the first 10 seconds of the motion. Explain how the system's performance has improved. Is this system overdamped, underdamped or critically damped?

4. A simple LRC electrical circuit consists of a capacitor with a capacitance of 0.02 farads, a resistor with a resistance of 40 ohms and an inductor with an inductance of 8 henrys. The circuit is connected to a 24-volt battery. Initially there is no charge on the capacitor and no current in the circuit.

Produce a mathematical model which gives the charge on the capacitor for any time after the switch is closed. Find the charge on the capacitor after 1 sec and the current in the circuit after 1 sec.

The LRC circuit is now connected to an alternating current source which applies a voltage

$$E(t) = 100 \cos 2t \quad \text{(in volts)}.$$

There is no initial charge on the capacitor or current in the circuit.

(a) Find an equation which gives the charge on the capacitor for any time after the switch is closed.

(b) Find the charge on the capacitor after 1 sec.

(c) Find the current in the circuit after 1 sec.

(d) Would a 5-ampere fuse have its capacity exceeded in this circuit?

(e) Use the computer to graph the transient response of the circuit (i.e., the complementary function of the differential equation which models the circuit).

(f) Use the computer to graph the general solution to the differential equation which models the circuit. Explain what happens to the transient response as time increases. Also, explain what happens to the steady state solution.

5. Consider the following economic model. Let P be the price of a single item on the market and Q be the quantity of the item available on the market. Both P and Q are functions of time t. By considering price and quantity as two interacting species, the following mathematical model can be proposed:

$$\frac{dP}{dt} = c_1 P \left(\frac{c_2}{Q} - P \right),$$

$$\frac{dQ}{dt} = c_3 Q (c_4 P - Q),$$

where c_1, c_2, c_3 and c_4 are positive constants. Justify and discuss the adequacy of this model.

(a) If $c_1 = 1$, $c_2 = 10,000$, $c_3 = 1$ and $c_4 = 25$, find the equilibrium points of this system. Classify each equilibrium point with respect to its stability. Give an explanation in cases where a point cannot be readily classified.

(b) Use the computer to perform a graphical stability analysis to determine what will happen to the levels of P and Q as time increases.

(c) Give an economic interpretation of the curves that determine the equilibrium points.

6. (a) For a simple *RL* circuit, Kirchhoff's voltage law requires that (if Ohm's law holds)

$$L\frac{dI}{dt} + RI = 0,$$

where L is inductance, R is resistance and I is current. Solve for I in the case $L = R = 2$ and $I(0) = 0.01$. Use both an analytical method and a numerical method.

(b) In contrast to part (a), real resistors may not always obey Ohm's law. For example, the voltage drop may be nonlinear and the circuit dynamics may be described by a relationship such as

$$L\frac{dI}{dt} + \left[\frac{-I}{I_{ref}} + \left(\frac{I}{I_{ref}}\right)^3\right]R = 0,$$

where all other parameters are as defined in (a) and I_{ref} is a known reference current equal to 1. Solve for I as a function of time under the same conditions as specified in (a).

7. Apart from inflow and outflow, another method by which mass can enter or leave a reactor is by a chemical reaction. For example, if the chemical decays, the reaction can sometimes be characterized as a first-order reaction, namely:

$$\text{Reaction} = -RVC,$$

where $V =$ volume (m^3), $c =$ concentration (moles/m^3) and $R =$ reaction rate (min^{-1}), which can generally be interpreted as the fraction of the chemical which goes away per unit time. So, if $R = 0.1\text{min}^{-1}$, for example, then approximately 10% of the chemical in the reactor decays in one minute. On substituting the reaction into the mass-balance equation, we have

$$V\frac{dc}{dt} = Fc_{in} - Fc - RVc,$$

where $F = $ flow rate (m^3 / min).

(a) Find the steady-state concentration of the reactor in the case where $R = 0.25\,min^{-1}$, $c_{in} = 50mg/min^3$, $F = 10m^3 / min$ and $V = 200m^3$.

(b) Repeat part (a), but compute the transient concentration response for the case $c_0 = 20mg/m^3$. Validate the results using Euler's numerical method from $t = 0$ to 30 min.

8. Biomedical and environmental engineers must frequently predict the outcome of predator-prey or host-parasite relationships. A simple model of such interacts is provided by the following system of ODE's:

$$\frac{dx_1}{dt} = g_1x_1 - d_1x_1x_2,$$

$$\frac{dx_2}{dt} = -d_2x_2 + g_2x_1x_2,$$

where x_1 and x_2 are the numbers of hosts and parasites, respectively. The d's and g's are death and growth rates, respectively, where the subscript 1 refers to the host and the 2 to the parasite. Notice that the deaths of the host and the growth of the parasite are dependent on both x_1 and x_2.

Use numerical methods to compute values of $x_1(t)$ and $x_2(t)$ from $t = 0$ to 10 for the following case:

$$x_1(0) = 10, \; x_2(0) = 2, \; g_1 = 1, \; d_1 = 0.2, \; g_2 = 0.01, \; d_2 = 0.5.$$

Use the computer to plot graphs of $x_1(t)$ and $x_2(t)$ against t. Interpret the results.

9. A population of 1,000,000 people is subject to a disease which is seldom fatal and leaves the victim immune to future infections by this disease. Infection can only occur when a susceptible person comes into direct contact with an infectious person. The infectious period lasts approximately four weeks. Last week there were 25 new cases of the disease reported. This week there were 48 new cases. It is estimated that 25% of the population is immune due to previous exposure.

(a) Develop a mathematical model as a discrete-time dynamical system. Hence find the eventual number of people who will become infected.

(b) Estimate the maximum number of new cases in any one week.

(c) Conduct a sensitivity analysis to investigate the effect of any assumptions made in part (a) which were not supported by hard data.

(d) Perform a sensitivity analysis for the number (25) of cases reported last week. It is thought by some that in early weeks the epidemic might be under-reported.

10. Consider a uniform beam of length l subject to a linearly increasing distributed load

$$W(x) = \frac{W_0 x}{l}, \quad 0 < x < l.$$

Assume the beam is hinged at the end $x = 0$ and imbedded at the end $x = l$.

By solving the governing ODE

$$\frac{d^4 y}{dx^4} = \frac{W(x)}{EI}, \quad 0 < x < l,$$

where E is Young's modulus of elasticity and I is the moment of inertia of the cross section about the neutral axis, show that the resulting deflection is given by

$$y = \frac{W_0}{120 EIl}(-x^5 + 2l^2 x^3 - l^4 x) \cdot$$

Taking the following parameter values:

$$l = 200\text{in}, \quad E = 29 \times 10^6 \text{lb/in}^2,$$

$$I = 725\text{in}^4, \quad W_0 = 100\text{lb/ft} \cdot$$

use the computer to plot the elastic curve. Also use a numerical method to determine the point of maximum deflection, expressing your result in inches.

Part II

PARTIAL DIFFERENTIAL

EQUATIONS

Case Study A3

CYLINDRICAL AND SPHERICAL SOLIDIFICATION IN HEAT TRANSFER

SUMMARY: This case study uses mathematical modelling to describe, develop and compare several effective methods for the numerical solution of one-dimensional Stefan problems. It is not intended to be an exhaustive treatment but is restricted to a range of problems and geometries including melting in the half-plane, outward cylindrical solidification and outward spherical solidification. The methods used include the enthalpy method, boundary immobilization method, perturbation method, nodal integral method and heat balance integral method. From the comparison of numerical results obtained the models can be validated and some helpful comments can be made which may prove valuable in the future use of these methods for problems of this type.

1. Background

Phase change problems, also known as Stefan problems, occur naturally in many physical processes, such as, freezing and thawing of foods, production of ice, ice formation on pipe surface, solidification of steel and chemical reaction. Mathematically, melting/solidification problems are special cases of moving boundary problems. Problems in which the solution of a differential equation has to satisfy conditions on the boundary of a prescribed domain are referred to as boundary-value problems. However, in the case of melting/solidification, the boundary of the domain is not known in advance. This means that the solution of such problems requires solving the diffusion or heat conduction equation in an unknown region which has to be determined as part of the solution.

There are very limited analytical solutions to melting/solidification problems and existing closed-form solutions to these significant problems are highly restrictive as to permissible initial and boundary conditions. So numerical solution becomes the main tool in the study of moving-boundary problems. Two conditions are required in order to solve these moving-boundary problems, one to determine the boundary itself and the other to complete the definition of the solution of the differential equation.

This case study involves a brief review of possible modelling and numerical approaches for one dimensional Stefan problems for simple geometries including plane, cylindrical and spherical. Numerical results are obtained from a range of methods researched by Caldwell and Kwan, including the enthalpy method, boundary immobilization method (BIM), perturbation method, nodal integral method (NIM) and heat balance integral method (HBIM). By comparing results, and in some cases making comparisons with analytical solutions (where possible), the models can be validated and some constructive comments can be made which will provide useful guidelines for the future use of these methods.

2. Problem Statement

First we consider the melting of material initially at its freezing temperature in the half-plane $x > 0$ subject to a time-dependent temperature change at $x = 0$. Then we extend this work to more realistic situations such as

(1) outward cylindrical solidification
(2) outward spherical solidification

of a saturated liquid due to low temperature at the boundary. Our aim is to calculate the position of the moving boundary for different times.

In order to gain confidence in the various models and methods we compare numerical results for test cases involving plane geometry, outward cylindrical and outward spherical solidification. For plane geometry there is an analytic solution but this does not apply to cylindrical and spherical geometries. Hence, it is important in these cases to validate results by comparing numerical results from a number of established methods to a number of test cases.

3. Model Formulation

3.1 MELTING IN THE HALF-PLANE

Consider the melting of certain material initially at its freezing temperature T_f in the half-plane $x > 0$ subject to a time-dependent temperature change at $x = 0$. The governing equation for the process is

$$\frac{\partial T}{\partial t} = \frac{\partial^2 T}{\partial x^2}, \; 0 < x < s(t), \; t > 0.$$
(A3.1)

subject to boundary conditions

$$T(x=0,t) = f(t), \; T(x=s(t),t) = 0,$$
(A3.2)

$$\frac{ds}{dt} = -\alpha \left(\frac{\partial T}{\partial x} \right)_{x=s(t)},$$
(A3.3)

where T is the temperature, x is the space variable, $s(t)$ is the position of the moving boundary and $\alpha = c(T_f - T_{ref})/L$ is the Stefan number, where c is the specific heat, L is the latent heat and T_{ref} is some reference temperature. For example, one can select T_{ref} such that $f(t=0) = 1$ or $\max_{0 \le t \le t_{final}} |f(t)| = 1$.

3.2 OUTWARD CYLINDRICAL SOLIDIFICATION

Consider the outward cylindrical solidification of a saturated liquid due to low temperature at the boundary. The problem can be formulated as

$$\frac{\partial T}{\partial t} = \frac{1}{r} \frac{\partial}{\partial r} \left(r \frac{\partial T}{\partial r} \right), \; 1 < r < s(t), \; t > 0,$$
(A3.4)

$$T(r=1,t) = f(t), \; T(r=s(t),t) = 1,$$
(A3.5)

$$\frac{ds}{dt} = \alpha \left(\frac{\partial T}{\partial r} \right)_{r=s(t)}.$$
(A3.6)

3.3 OUTWARD SPHERICAL SOLIDIFICATION

In the case of outward spherical solidification, the corresponding governing equation is

$$\frac{\partial T}{\partial t} = \frac{1}{r} \frac{\partial^2 (rT)}{\partial r^2}, \; 1 < r < s(t), \; t > 0,$$
(A3.7)

subject to boundary conditions (A3.5) and (A3.6).

4. Mathematical/Numerical Solution

In this section we introduce a range of possible methods for the solution of Stefan problems of the type in question. Except for the heat balance integral method, where the formulation for cylindrical geometry is introduced, the other methods only include the formulation for the plane geometry since other applications will follow the same idea. Readers may refer to specific papers listed in the References for further details of the methods.

4.1 ENTHALPY METHOD

The enthalpy formulation is one of the most popular fixed-domain methods for solving the Stefan problem. In the formulation, the enthalpy function is introduced such that the flux condition is automatically satisfied across the phase front, which is realized as a jump discontinuity of the enthalpy. Date [12] has developed an enthalpy method which tracks the phase front easily. He has applied this method to one and two dimensional problems in plane geometry and has obtained good agreement with existing solutions. More recently, Caldwell et al. [2, 3] have also successfully applied the method to cylindrical and spherical geometries.

First, the enthalpy function H is defined by

$$H = T + \alpha' f_l(T),\qquad\qquad(A3.8)$$

where $\alpha' = \dfrac{1}{\alpha}$ and f_l is the local liquid fraction given by

$$f_l(T) = \begin{cases} 1 & \text{if } T \geq 1, \\ 0 & \text{if } T < 0. \end{cases}$$

Hence H is identical to the temperature except when phase change occurs, in which case H has a jump of α'. Substituting H into the heat equation, we obtain

$$\frac{\partial H}{\partial t} = \frac{\partial^2 T}{\partial x^2}.\qquad\qquad(A3.9)$$

Discretization of (A3.9) will result in a set of non-linear equations. Date [12] introduces a simple method which at the same time provides an effective means of tracking the phase boundary. From (A3.8) we can write $T = H + H'$, where

$$H' = -\alpha' f_l(T) = \begin{cases} -\alpha' & \text{if } H \geq \alpha', \\ -H & \text{if } 0 < H < \alpha', \\ 0 & \text{if } H \leq 0. \end{cases}\qquad(A3.10)$$

Also, we note that $-H'/\alpha'$ is the local liquid fraction while $1+H'/\alpha'$ is the local solid fraction.

The implicit discretization of (A3.9) is

$$\frac{H_i^{(k+1)} - H_i^{(k)}}{\Delta t} = \frac{T_{i-1}^{(k+1)} - 2T_i^{(k+1)} + T_{i+1}^{(k+1)}}{(\Delta x)^2}, \; i=1,2,\cdots,N-1, \qquad (A3.11)$$

where Δx and Δt represent the space and time steps, respectively. Using the relation $T_i^{(k)} = H_i^{(k)} + H_i'^{(k)}$ with $H_i'^{(k)}$ obtained from (A3.10), we have

$$-\gamma H_{i-1}^{(k+1)} + (1+2\gamma)H_i^{(k+1)} - \gamma H_{i+1}^{(k+1)} = H_i^{(k)} + \gamma(H_{i-1}'^{(k+1)} - 2H_i'^{(k+1)} + H_{i+1}'^{(k+1)}), \quad (A3.12)$$

where $\gamma = \Delta t /(\Delta x)^2$. This results in a set of nonlinear equations. To solve this system we employ an iterative scheme, where terms involving H' are set to lag behind terms involving H for one iteration. Using the value of H from the previous time step as the initial guess, the values of H' are calculated from (A3.10). The new value of H is then obtained from (A3.12). This process is continued until the iterations converge. Then we can continue to the next time step. Note that each iteration involves solving a tridiagonal system, and can be done effectively by the Thomas algorithm.

Recalling that $-H_i'/\alpha'$ is the liquid fraction in the ith control volume, there is a simple way to calculate the position of the phase front. Consider the integral I, which represents the volume of solid in the range $0 \le x \le s(t)$:

$$I = \int_0^{s(t)} dz \approx \Delta x \sum_{i=1}^{N}\left(-\frac{H_i'}{\alpha'}\right) + \frac{\Delta x}{2}.$$

The last term is due to the fact that the first cell is always occupied by liquid. Here, the summation can be carried out over all the cells since the cells behind the phase-front give zeros to $-H_i'/\alpha'$. Hence we have $s(t) = I$.

Note that in the cases of outward cylindrical and spherical solidification, there are small differences in the formulae for I and $s(t)$. The corresponding equations for outward cylindrical solidification are

$$I = \int_1^{s(t)} z\,dz = \Delta z \sum_{i=1}^{N} z_i \left(1+\frac{H_i'}{\alpha'}\right) + \frac{\Delta z}{2},$$

$$s(t) = \sqrt{1+2I}$$

and the equations for outward spherical solidification are

$$I = \int_1^{s(t)} z^2 dz = \Delta z \sum_{i=1}^{N} z_i^2 \left(1 + \frac{H_i'}{\alpha'}\right) + \frac{\Delta z}{2},$$

$$s(t) = \sqrt[3]{1 + 3I}.$$

4.2 BOUNDARY IMMOBILIZATION METHOD (BIM)

With a suitable transformation, it is possible to fix the moving boundary. This method was first applied to a finite difference scheme by Crank [11]. Kutluay et al. [16] have also successfully applied the method to various problems.

Under the transformation

$$x^* = \frac{x}{s}, \ T^*(x^*,t) = T(x,t),$$

the problem (A3.1)-(A3.3) can be transformed to one in the fixed domain $0 \le r' \le 1$:

$$s^2 \frac{\partial T^*}{\partial t} = \frac{\partial^2 T^*}{\partial x^{*2}} + x^* s \frac{ds}{dt} \frac{\partial T^*}{\partial x^*} \tag{A3.13}$$

subject to

$$T^*(x^* = 0, t) = f(t), \ T^*(x^* = 1, t) = 0, \tag{A3.14}$$

$$s \frac{ds}{dt} = -\alpha \left(\frac{\partial T^*}{\partial x^*}\right)_{x^*=1}. \tag{A3.15}$$

A finite difference discretization of (A3.13) implicit in T^* and explicit in s is

$$a_i^{(k+1)} T_{i-1}^{(k+1)} + b_i^{(k+1)} T_i^{(k+1)} + c_i^{(k+1)} T_{i+1}^{(k+1)} = (s^{(k)})^2 T_i^{(k)}, \tag{A3.16}$$

where

$$a_i^{(k+1)} = \gamma \left[\frac{\Delta x}{2} x_i s^{(k)} \left(\frac{ds}{dt}\right)^{(k)} - 1\right],$$

$$b_i^{(k+1)} = (s^{(k)})^2 + 2\gamma,$$

$$c_i^{(k+1)} = -a_i^{(k+1)} - 2\gamma,$$

$$\left(\frac{ds}{dt}\right)^{(k)} = \frac{\alpha}{s^{(k)}}(4T_{N-1}^{(k)} - T_{N-2}^{(k)}),$$ (A3.17)

and $\gamma = \Delta t/(\Delta x)^2$. At each time step, the temperature distribution is obtained by solving the tridiagonal system (A3.16) and the position of the moving boundary is updated via the formula

$$s^{(k+1)} = s^{(k)} + \left(\frac{ds}{dt}\right)^{(k)} \Delta t.$$ (A3.18)

Note that a starting solution for small time is required by the BIM. For plane geometry one can use the analytic solution for the problem with constant boundary condition as the starting solution (see Caldwell and Savovic [4]). Readers may refer to Caldwell and Kwan [5] for the starting solutions for other geometries. More recently, the finite-difference/BIM approach has been extended to deal with the Stefan problem with periodic boundary conditions (see Savovic and Caldwell [19]).

4.3 PERTURBATION METHOD

The perturbation method only works for small Stefan number. It has been successfully applied to Stefan problems with simple boundary conditions in different geometries; see Huang and Shih [15], Pedroso and Domoto [17], Stephan and Holzknecht [20]. More recently, Caldwell and Kwan [6] successfully applied the method to Stefan problems with time-dependent boundary conditions. Further benchmark cases are presented by Caldwell et al. [8] showing a high degree of agreement and accuracy when nodal integral and finite-difference solutions are compared with exact solutions.

Since $s(t)$ is expected to be a monotonic function of t, we may replace t by s as the second independent variable in the governing equations. By making use of (A3.3), (A3.1) can be written as

$$\frac{\partial^2 T}{\partial x^2} = -\alpha \frac{\partial T}{\partial s}\left(\frac{\partial T}{\partial x}\right)_{x=s}.$$ (A3.19)

On the other hand, the boundary condition at $x = 0$ is written as

$$T = f(t) = F(s) \text{ on } x = 0.$$ (A3.20)

We now derive a three term perturbation solution of the form

$$T(x,s) = T_0(x,s) + \alpha T_1(x,s) + \alpha^2 T_2(x,s). \tag{A3.21}$$

Substituting (A3.21) into (A3.19) and (A3.20), the governing equations for T_0, T_1 and T_2 are

$$\alpha^0: \quad \frac{\partial^2 T_0}{\partial x^2} = 0,$$
$$T_0(x=0,s) = F(s), \quad T_0(x=s,s) = 0.$$

$$\alpha: \quad \frac{\partial^2 T_1}{\partial x^2} = -\frac{\partial T_0}{\partial s}\left(\frac{\partial T_0}{\partial x}\right)_{x=y},$$
$$T_1(x=0,s) = 0, \quad T_1(x=s,s) = 0.$$

$$\alpha^2: \quad \frac{\partial^2 T_2}{\partial x^2} = -\frac{\partial T_0}{\partial s}\left(\frac{\partial T_i}{\partial x}\right)_{x=s} - \frac{\partial T_i}{\partial s}\left(\frac{\partial T_0}{\partial x}\right)_{x=s},$$
$$T_2(x=0,s) = 0, \quad T_2(x=s,s) = 0. \tag{A3.22}$$

The solutions of the above equations are

$$T_0(x,s) = F(s)(1-z),$$
$$T_1(x,s) = \frac{1}{6}F(s)z(z-1)[F(s)(z+1) - F'(s)s(z-2)],$$
$$T_2(x,s) = \frac{-1}{360}F(s)z(z-1)[F(s)^2(z+1)(9z^2+19) + 10F'(s)^2 y^2(z+4)$$
$$+5F(s)F'(s)s(3z^2 + 5z + 17)$$
$$+F(s)F''(s)s^2(z-2)(3z^2 - 6z - 4)], \tag{A3.23}$$

where $z = x/s$. Thus, the position of the moving boundary follows the equation

$$\frac{ds}{dt} = -\alpha\left(\frac{\partial T_0}{\partial x} + \alpha\frac{\partial T_1}{\partial x} + \alpha^2\frac{\partial T_2}{\partial x}\right)_{x=s}$$
$$= \frac{\alpha}{s}F(s) - \alpha^2 F(s)\left[\frac{1}{6}F'(s) + \frac{1}{3s}\right]$$
$$+ \alpha^3 F(s)\left[\frac{7}{45s}F(s)^2 + \frac{5}{36}F'(s)^2 s + \frac{25}{72}F(s)F'(s) - \frac{13}{360}F(s)F''(s)s\right]. \tag{A3.24}$$

The final step is to substitute back $f(t)$ for $F(s)$. With the relations

$$\frac{dF(s)}{ds} = \frac{df(t)}{dt}\left(\frac{ds}{dt}\right)^{-1}, \quad \frac{d^2F(s)}{ds^2} = \frac{d^2f(t)}{dt^2}\left(\frac{ds}{dt}\right)^{-2},$$

(A3.24) can be rewritten in the form

$$\left(\frac{ds}{dt}\right)^3 + a(t,s)\left(\frac{ds}{dt}\right)^2 + b(t.s)\frac{ds}{dt} + c(t,s) = 0, \qquad \text{(A3.25)}$$

where

$$a(t,s) = -\frac{\alpha f(t)}{s}\left[1 - \frac{\alpha}{3}f(t) + \frac{7\alpha^2}{45}f(t)^2\right],$$

$$b(t,s) = \alpha^2 f(t)f'(t)\left[\frac{1}{6} - \frac{25\alpha}{72}f(t)\right],$$

$$c(t,s) = -\alpha^3 f(t)s\left[\frac{5}{36}f'(t)^2 - \frac{13}{360}f(t)f''(t)\right].$$

By solving the cubic equation (A3.25), the value of $\frac{ds}{dt}$ is obtained and s can be found by numerical integration. On the other hand, the temperature distribution can be obtained by substituting (A3.16) into (A3.14).

4.4 NODAL INTEGRAL METHOD (NIM)

A semianalytical nodal method to solve the one-dimensional Stefan problem was recently developed by Rizwan-uddin [18]. We give a brief description of the method here. The space-time domain ($0 \le x \le 1$; $0 \le t \le t_{final}$) is first discretized into space-time nodes. Each node is identified by the subscript (i,j). The space-averaged, time-dependent temperature and time-averaged, space-dependent temperature for each node are respectively defined as

$$\overline{T}_i^x(t) \equiv \frac{1}{\Delta x}\int_{x_i}^{x_i+\Delta x} T(x,t)\,dt, \quad \overline{T}_j^t(x) \equiv \frac{1}{\Delta t}\int_{t_{j-1}}^{t_j} T(x,t)\,dt.$$

First, (A3.15) is integrated over the time step $t_{j-1} \le t \le t_j$ to yield

$$s^2(t) = s^2(t_{j-1}) - 2Ste \frac{d\overline{T}_j^t(x=1)}{dx}(t-t_{j-1}), \ t_{j-1} \le t \le t_j.$$

Next, for each space-time node, a time-step-averaged, second-order ODE is obtained for $\overline{T}_j^t(x)$ by operating on (A3.13) with $\frac{1}{\Delta t} \int_{t_{j-1}}^{t_j} dt$, and a space-averaged, first-order ODE for $\overline{T}_i^x(t)$ is obtained by operating on (A3.13) with $\frac{1}{\Delta x} \int_{x_i}^{x_i+\Delta x} dx$. After introducing some simplifying assumptions, the second-order ODE in space is solved using the Dirichlet boundary conditions at the left and right edge of the node, leading to a solution of the form

$$\overline{T}_j^t(x) = C_1 + C_2 x + g(e^{-C_3 x}),$$

where C_m ($m = 1, 2, 3, \cdots$) are constants. On the other hand, the first-order ODE in time is solved using the initial condition at the beginning of the time step, leading to a solution of the form

$$\overline{T}_i^x(t) = \overline{T}_i^x(t_{j-1}) + C_4 \ln[1 - C_5(t - t_{j-1})].$$

Further details of the method and its application to two problems with time-dependent boundary conditions are given by Caldwell and Kwan [7]. A comparison of numerical results with those from the enthalpy method is also included. Benchmark cases are presented by Caldwell et al. [8] involving two test examples with the aim of producing very high accuracy when validated against the exact solutions.

4.5 HEAT BALANCE INTEGRAL METHOD (HBIM)

The heat balance integral method was first proposed by Goodman [13, 14]. Goodman's idea is to assume a particular temperature profile, and then integrate the heat equation over an appropriate interval to obtain a set of heat-balance integral equations. The equations are then solved to obtain the motion of the phase boundary. Later Bell [1] proposed a systematic method to improve the accuracy of HBIM, which we will present below. The main idea is to subdivide the dependent variable T, and assume a linear profile within each subdivision. Accuracy can be improved by increasing the number of subdivisions. A detailed description of the method can also be found in Caldwell and Chiu [9].

The method is as follows: First we divide the range $[0,1]$ into N parts, that is

$$T_i = \frac{i}{N},$$

and denote the corresponding position of the isotherm by Z_i. Assume a linear profile within each subdivision $[Z_i, Z_{i+1}]$,

$$T(r) = \frac{i}{N} + \frac{r - Z_i}{N(Z_{i+1} - Z_i)} \quad \text{for } Z_i \le r \le Z_{i+1}. \tag{A3.26}$$

Multiplying (A3.4) by r and integrating over $[Z_i, Z_{i+1}]$ gives

$$\int_{Z_i}^{Z_{i+1}} r \frac{\partial T}{\partial t} dr = \int_{Z_i}^{Z_{i+1}} \frac{\partial}{\partial r}\left(r \frac{\partial T}{\partial r} \right) dr \cdot$$

Taking the derivative outside the integral sign, we obtain

$$\frac{d}{dt}\left(\int_{Z_i}^{Z_{i+1}} rT\, dr - \frac{Z_{i+1}^2 T_{i+1}}{2} + \frac{Z_i^2 T_i}{2} \right) = \left(r \frac{\partial T}{\partial r} \right)_{Z_{i+1}} - \left(r \frac{\partial T}{\partial r} \right)_{Z_i} \cdot$$

Replacing T by the linear profile and ensuring that the expression representing change in flux is approximated by the discontinuous change in adjacent profile gradients, we obtain a system of ordinary differential equations for the penetration depth Z_i, namely,

$$(2Z_1 + 1)\dot{Z}_1 = \frac{6}{Z_1 - 1} - \frac{6Z_1}{Z_2 - Z_1},$$

$$(2Z_{i+1} + Z_i)\dot{Z}_{i+1} + (Z_{i+1} + 2Z_i)\dot{Z}_i = \frac{6Z_i}{Z_{i+1} - Z_i} - \frac{6Z_{i+1}}{Z_{i+2} - Z_{i+1}}, \quad i = 1, 2, \cdots, N-2,$$

$$[2(1 + 3N/\alpha)Z_N + Z_{N-1}]\dot{Z}_N + (Z_N + 2Z_{N-1})\dot{Z}_{N-1} = \frac{6Z_{N-1}}{Z_N - Z_{N-1}}. \tag{A3.27}$$

It can be seen that the system (A3.27) is stiff, at least for small t where the distances between adjacent isotherms are small. Hence a stiff ODE solver is required to solve the system. Besides, a starting solution is required by the HBIM. Readers may refer to Caldwell and Chiu [10] for a choice of the starting solution. An alternative choice would be the one mentioned in the BIM.

5. Model Validation

In this section we present and discuss the numerical results in applying the above methods to different test cases.

5.1 TEST EXAMPLE 1

The first example corresponds to the melting in plane geometry with $\alpha = 0.2$ and $f(t) = 1$. The analytic solution to the problem is

$$T(x,t) = 1 - \frac{erf(x/(2\sqrt{t}))}{erf(\lambda)}, \ s(t) = 2\lambda\sqrt{t}, \tag{A3.28}$$

where *erf* denotes the error function and λ is the solution of the transcendental equation

$$\sqrt{\pi}\lambda\exp(\lambda^2)erf(\lambda) = \alpha. \tag{A3.29}$$

The numerical results for this example are presented in Table A3.1. Note that an adaptive ODE solver is used in the perturbation method and so the time step is not constant.

5.2 TEST EXAMPLE 2

The second example corresponds to the melting in plane geometry with $\alpha = 1.0$ and $f(t) = \exp(t) - 1$. The analytic solution to the problem is

$$T(x,t) = \exp(t - x) - 1, \ s(t) = t. \tag{A3.30}$$

The numerical results for this example are presented in Table A3.2.

5.3 TEST EXAMPLE 3

This example corresponds to the outward cylindrical solidification with $\alpha = 0.2$ and $f(t) = 0$. There is no known analytical solution to the problem. The numerical results for this example are presented in Table A3.3. Note that the adaptive ODE solver is also used in the HBIM.

5.4 TEST EXAMPLE 4

The last example corresponds to the outward spherical solidification with $\alpha = 0.2$ and $f(t) = 0$. There is also no known analytical solution to the problem. The numerical results for this example are presented in Table A3.4.

5.5 DISCUSSION

In the case of plane melting, as reflected in Tables A3.1 and A3.2, the methods employed give good results in predicting the position of the moving boundary when compared with the analytic solutions. In the cases of cylindrical and spherical solidification, where the analytic solutions are not available, the methods employed give very similar results, as reflected in Tables A3.3 and A3.4. The good agreement achieved gives us confidence in using the methods to solve the Stefan problem numerically for different geometries.

6. Interpretation and Conclusions

Here we give some general comments on the methods, which can serve as a guideline for solving a particular Stefan problem.

The enthalpy method is popular because of its easy formulation. As the governing equation for the enthalpy is very similar to that for temperature, only little extra effort is required in programming. However, the iterative nature of the solution procedure makes the computational time longer. Besides, normally the enthalpy method produces an unphysical oscillating solution near the moving boundary. The extension of the method to higher-dimensional problems is also difficult due to the lack of efficient methods of locating the moving boundary.

Table A3.1. Melting in plane geometry ($\alpha = 0.2$, $f(t) = 1$).

Time	Exact	Enthalpy	BIM	Perturbation	NIM
0.100	0.19380	0.19400	0.19433	0.19386	0.19382
0.200	0.27407	0.27425	0.27456	0.27416	0.27410
0.300	0.33567	0.33572	0.33612	0.33578	0.33571
0.400	0.38759	0.38756	0.38802	0.38772	0.38764
0.500	0.43334	0.43345	0.43375	0.43349	0.43340
0.600	0.47470	0.47487	0.47509	0.47486	0.47476
0.700	0.51274	0.51281	0.51311	0.51291	0.51280
0.800	0.54814	0.54811	0.54850	0.54832	0.54821
0.900	0.58139	0.58141	0.58174	0.58158	0.58146
1.000	0.61284	0.61291	0.61318	0.61304	0.61292
1.100	0.64275	0.64282	0.64309	0.64297	0.64283
1.200	0.67133	0.67135	0.67166	0.67156	0.67142
1.300	0.69875	0.69872	0.69906	0.69898	0.69883
1.400	0.72512	0.72520	0.72543	0.72536	0.72521
1.500	0.75057	0.75057	0.75088	0.75082	0.75067
1.600	0.77519	0.77526	0.77549	0.77545	0.77529
1.700	0.79905	0.79903	0.79934	0.79931	0.79915
1.800	0.82221	0.82225	0.82250	0.82248	0.82232
1.900	0.84474	0.84487	0.84503	0.84502	0.84485
2.000	0.86669	0.86668	0.86697	0.86697	0.86680
N		100	100	NA	8
Δt		0.001	0.001	NA	0.01

Table A3.2. Melting in plane geometry ($\alpha = 1.0$, $f(t) = \exp(t) - 1$).

Time	Exact	Enthalpy	BIM	NIM
0.050	0.05000	0.05053	0.05001	0.05000
0.100	0.10000	0.10053	0.10002	0.10000
0.150	0.15000	0.15052	0.15002	0.15000
0.200	0.20000	0.20053	0.20003	0.20000
0.250	0.25000	0.25054	0.25004	0.25000
0.300	0.30000	0.30055	0.30005	0.29999
0.350	0.35000	0.35056	0.35005	0.34999
0.400	0.40000	0.40057	0.40006	0.39998
0.450	0.45000	0.45058	0.45007	0.44998
0.500	0.50000	0.50059	0.50008	0.49997
0.550	0.55000	0.55061	0.55008	0.54996
0.600	0.60000	0.60062	0.60009	0.59994
0.650	0.65000	0.65063	0.65010	0.64993
0.700	0.70000	0.70064	0.70010	0.69991
0.750	0.75000	0.75066	0.75011	0.74989
0.800	0.80000	0.80067	0.80011	0.79986
0.850	0.85000	0.85068	0.85012	0.84984
0.900	0.90000	0.90070	0.90013	0.89981
0.950	0.95000	0.95071	0.95013	0.94977
1.000	1.00000	1.00072	1.00014	0.99974
N		100	100	8
Δt		0.001	0.001	0.01

Table A3.3. Outward cylindrical solidification ($\alpha = 0.2$, $f(t) = 0$).

Time	Enthalpy	BIM	Perturbation	HBIM
0.100	1.18850	1.18989	1.18882	1.18824
0.200	1.26415	1.26504	1.26421	1.26349
0.300	1.32101	1.32190	1.32120	1.32032
0.400	1.36847	1.36932	1.36869	1.36766
0.500	1.40997	1.41072	1.41016	1.40898
0.600	1.44715	1.44786	1.44735	1.44603
0.700	1.48117	1.48179	1.48132	1.47986
0.800	1.51264	1.51318	1.51274	1.51114
0.900	1.54197	1.54251	1.54209	1.54036
1.000	1.56954	1.57010	1.56971	1.56785
1.100	1.59573	1.59623	1.59586	1.59387
1.200	1.62058	1.62108	1.62073	1.61863
1.300	1.64446	1.64483	1.64449	1.64227
1.400	1.66711	1.66759	1.66727	1.66492
1.500	1.68901	1.68947	1.68917	1.68670
1.600	1.71013	1.71057	1.71027	1.70770
1.700	1.73053	1.73095	1.73066	1.72798
1.800	1.75027	1.75067	1.75040	1.74760
1.900	1.76940	1.76980	1.76954	1.76664
2.000	1.78799	1.78839	1.78814	1.78512
N	100	100	NA	32
Δt	0.001	0.001	NA	NA

Table A3.4. Outward spherical solidification ($\alpha = 0.2$, $f(t) = 0$).

Time	Enthalpy	BIM	Perturbation	HBIM
0.100	1.18368	1.18499	1.18375	1.18309
0.200	1.25469	1.25564	1.25469	1.25378
0.300	1.30732	1.30831	1.30751	1.30635
0.400	1.35092	1.35176	1.35105	1.34965
0.500	1.38856	1.38936	1.38872	1.38709
0.600	1.42217	1.42284	1.42225	1.42041
0.700	1.45261	1.45321	1.45267	1.45062
0.800	1.48054	1.48115	1.48064	1.47839
0.900	1.50652	1.50711	1.50662	1.50418
1.000	1.53087	1.53142	1.53096	1.52833
1.100	1.55389	1.55432	1.55389	1.55108
1.200	1.57556	1.57603	1.57561	1.57262
1.300	1.59620	1.59668	1.59627	1.59312
1.400	1.61595	1.61640	1.61601	1.61268
1.500	1.63496	1.63529	1.63492	1.63143
1.600	1.65308	1.65345	1.65308	1.64943
1.700	1.67052	1.67093	1.67058	1.66676
1.800	1.68739	1.68780	1.68746	1.68349
1.900	1.70380	1.70411	1.70378	1.69966
2.000	1.71954	1.71991	1.71959	1.71532
N	100	100	NA	32
Δt	0.001	0.001	NA	NA

The BIM can effectively remove the moving nature of the boundary at the expense of solving a more complicated equation. Besides, a starting solution is required in order that the method can be started. One benefit of the BIM is that the computation time is comparatively short and so it is possible to achieve higher accuracy by refining the mesh size.

The perturbation method can transform the Stefan problem into an ODE for the position of the boundary. However, the formulation of the perturbation method is complicated and cannot be done easily without symbolic mathematical packages. Besides, the perturbation method only works for small Stefan numbers. Since α can be arbitrarily small by selecting T_{ref} close enough to T_f, a constraint on $f(t)$ is also required. Experience suggests that requiring $\max\limits_{0 \le t \le t_{final}} |f(t)| \le 1$ is a good constraint. In the case $f(t) = 1$, it is found that by adding more terms in the perturbation solution the method can work well for Stefan numbers as large as around 0.7.

The NIM can produce better results with comparatively small numbers of intervals. However, as the number of intervals increases the iteration will become more and more difficult to converge. Also the extension of the method to other geometries is difficult.

The HBIM gives good results for problems with constant boundary conditions. However, the extension to time-dependent problems is difficult. The complicated formulation also makes it less attractive. For these reasons the HBIM is normally used for validation purposes.

7. Computer Algorithms

Computer software has been developed for the five methods described in Section 4 but is too extensive to include here.

8. References and Bibliography

1. Bell, G.E., "A refinement of the heat balance integral method applied to a melting problem," International Journal of Heat and Mass Transfer, **21**, 1978, 1357-1362.
2. Caldwell, J. and Chan C.C., "Numerical solution of Stefan problems in annuli," in *Advanced Computational Methods in Heat Transfer VI*, (Eds: Sunden, B. and Brebbia, C.A.), WIT Press, Southampton and Boston, 2000, 215-225.
3. Caldwell, J. and Kwan, Y.Y., "Spherical solidification by the enthalpy method and heat balance integral method," in *Advanced Computational Methods in Heat Transfer VII*, (Eds: Sunden, B. and Breddia, C.A.), WIT Press, Southampton and Boston, 2002, 165-174.
4. Caldwell, J. and Savovic, S., "Numerical solution of Stefan problem by variable space grid and boundary immobilization method," Journal of Mathematical Sciences, **13**, 2002, 67-79.
5. Caldwell, J. and Kwan, Y.Y., "Small time solution for Stefan problems," submitted to Communication in Numerical Methods in Engineering.
6. Caldwell, J. and Kwan, Y.Y., "Perturbation methods for the Stefan problem with time-dependent boundary conditions," International Journal of Heat and Mass Transfer, **46**, 2003, 1497-1501.
7. Caldwell, J. and Kwan, Y.Y., "Nodal integral and enthalpy solution of one-dimensional Stefan problem," Journal of Mathematical Sciences, **13(2)**, 2002, 99-109.
8. Caldwell, J., Savovic, S. and Kwan, Y.Y., "Nodal integral and finite difference solution of one-dimensional Stefan problem," accepted for publication in Journal of Heat Transfer (ASME).
9. Caldwell, J. and Chiu, C.K., "Numerical solution of one-phase Stefan problems by the heat balance integral method, Part I – cylindrical and spherical geometries," Communications in Numerical Methods in Engineering, **16**, 2000, 569-583.
10. Caldwell, J. and Chiu, C.K., "Numerical solution of one-phase Stefan problems by the heat balance integral method, Part II – special small time starting procedure," Communications in Numerical Methods in Engineering, **16**, 2000, 585-593.

11. Crank, J., "Two methods for the numerical solution of moving boundary problems in diffusion and heat flow," Quarterly Journal of Mechanics and Applied Mathematics, **10(2)**, 1957, 220-231.
12. Date, A.W., "Novel strongly implicit enthalpy formulation for multi-dimensional Stefan problems," Numerical Heat Transfer B, **21**, 1992, 231-251.
13. Goodman, T.R., "The heat-balance integral and its application to problems involving a change of phase," Transactions of the ASME, **80**, 1958, 335-342.
14. Goodman, T.R., "The heat balance integral – further considerations and refinements," Journal of Heat Transfer, **83**, 1961, 83-86.
15. Huang, C.L. and Shih, Y.P., "Shorter Communications: perturbation solution for planar solidification of a saturated liquid with convection at the wall," International Journal of Heat and Mass Transfer, **18**, 1975, 1481-1483.
16. Kutluay, B., Bahadir, A.R. and Özdes, A., "The numerical solution of one-phase classical Stefan problem," Journal of Computational and Applied Mathematics, **81**, 1997, 135-144.
17. Perdroso, R.I. and Domoto, G.A., "Perturbation solutions for spherical solidification of saturated liquids," Journal of Heat Transfer, **95**, 1973, 42-46.
18. Rizwan-uddin, "A nodal method for phase change moving boundary problems," International Journal of Computational Fluid Dynamics, **11**, 1999, 211-221.
19. Savovic, S. and Caldwell, J., "Finite-difference solution of one-dimensional Stefan problem with periodic boundary conditions," accepted for publication in International Journal of Heat and Mass Transfer.
20. Stephan, K. and Holzknecht, B., "Perturbation solutions for solidification problems," International Journal of Heat and Mass Transfer, **19**, 1976, 597-602.

Case Study A4

ELASTIC ANALYSIS OF A SQUARE PLATE WITH CIRCULAR HOLES

SUMMARY: The problem under consideration involves the elastic analysis of a square plate subjected to a uniform pressure. In the first instance a square section with a central hole is considered using generalized plane strain element. The pressure is applied to the surface of a circular hole located at the centre of the section. This problem is then generalized to that of a square section with nine holes subjected to internal pressure. Results involving stresses and displacements are obtained for both cases using the BEASY Boundary Element software package from Computational Mechanics Ltd. A check on the accuracy is obtained by using Lamé thick cylinder theory at selected points. The good agreement obtained gives confidence in the use of the boundary element method for problems of this type.

1. Background

Elastic analysis of plates with various geometries, and in some cases containing holes, is important in many traditional engineering industries. Even in the food manufacturing industry it is important to be able to perform an elastic analysis of a plate with circular holes/cavities. In some cases the cavities may be representative of the structure of food itself. This means that one should be able to deal with the simplified problem when pressure is applied to the surface of a circular hole located at the centre of the section. An obvious extension is to both orderly and randomly arranged holes/cavities with pressure/shear. A typical application in the food industry would be to ice-cream boxes which contain holes. It is important to be able to calculate hoop stresses on cone stems to prevent damage in transit.

For problems of this type boundary element methods have advantages over finite element methods. For this reason the Boundary Element Analysis System (BEASY) software package from Computational Mechanics Ltd. has been chosen to carry out the analysis. Two useful references for this work are the book by Brebbia [2] which discusses the mathematical foundations of the work and a review paper by George [4] which discusses the relative merits of BE analysis and FE methods.

The boundary element method (BEM) has been successfully used in the solution of problems in linear elastostatic stress analysis. This case study outlines the basic underlying theory together with relevant analysis associated with a hollow cylinder subjected to uniform pressures on the inner and outer surfaces.

Results are obtained for the case of a square plate with a central hole subjected to internal pressure and generalized to the case of a square plate with nine holes. The stresses and displacements are computed by using the BEASY software package and validated in certain cases by using Lamé thick cylinder theory. Certain conclusions are drawn which include the advantages of the BEM over the FEM for problems of this type.

2. Problem Statement

At the first stage the geometry under consideration is a square section with a single central hole. An elastic analysis is required when the plate is subjected to a uniform pressure. The pressure is applied to the surface of a circular hole located at the centre of the section.

In the model only one quarter of the section need be considered due to symmetry (see Figure A4.1). It is assumed that the material behaves elastically with a Young's modulus of $50 \times 10^4 \, \text{lb/in}^2$ and Poisson's ratio of 0.2. The thickness of the section is taken to be 1 in and a uniform pressure of $1000 \, \text{lb/in}^2$ is applied to the inner surface of the hole. For the boundary conditions, zero displacements are assumed to exist on the lines of symmetry, i.e.,

$$u = 0 \ \text{at} \ x = 0,$$
$$v = 0 \ \text{at} \ y = 0.$$

At the second stage the geometry under consideration is generalized to one which contains 9 holes placed symmetrically in the square section. Again, in the model only one-quarter of the section need be considered due to symmetry (see Figure A4.2).

3. Model Formulation

3.1 BOUNDARY ELEMENT METHOD FOR ELASTOSTATIC PROBLEMS

The BEM is an ideal tool for the solution of problems in linear elastostatic stress analysis. The BEASY software package can be used to estimate the stresses and displacements which occur in a body subjected to a prescribed loading.

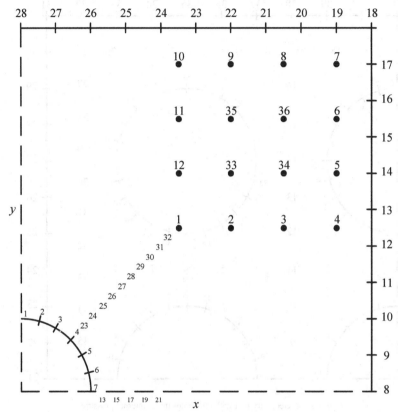

Figure A4.1. Geometry plot with defined elements for square section with a hole under internal pressure.

The principle of virtual displacements for linear elastic analysis of a problem domain V bounded by the boundary S may be written

$$\int_V \left(\sigma_{jk,j} + b_k\right) u_k^* \, dV = \int_{S_2} \left(p_k - \bar{p}_k\right) u_k^* \, dS + \int_{S_1} \left(\bar{u}_k - u_k\right) p_k^* \, dS \,,$$

in which the boundary 'tractions' (or pressures) p satisfy the boundary condition $p = \bar{p}$ on the portion S_2 of the problem boundary S, and the displacements u satisfy the boundary condition $u = \bar{u}$ on the portion S_1 of S. The term $\sigma_{jk,j}$ represent the derivatives of the stress tensor and b_k are the acting body forces. Also

u_k^* and p_k^* are the displacements and tractions corresponding to the virtual system.

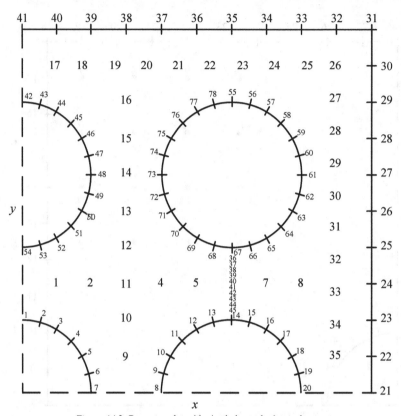

Figure A4.2. Geometry plot with nine holes under internal pressure.

Integrating this equation by parts twice and applying a 'fundamental' solution of the type

$$\sigma_{jk,j}^* + \Delta_1^i = 0 ,$$

where Δ_1^i is the Dirac delta function representing a unit load at 'i' in the '1' direction, we obtain

$$u_1^i + \int_S u_k p_k^* \ dS = \int_S p_k u_k^* \ dS + \int_V b_k u_k^* \ dV . \tag{A4.1}$$

In the absence of body forces b_k, this equation represents the boundary integral formulation for elastostatic analysis. However, the final term in equation (A4.1) may be transformed to an integral over the boundary S enabling problems involving rotational or gravitational loads to be solved.

Equation (A4.1) consists of a set of n equations and containing $2n$ unknowns (the traction and displacement in each direction at each node). However, in practice either the traction or the displacement (or some relationship between the two) is known at every node point and the number of unknowns may be reduced to n, enabling the equations to be solved.

For problems of this type the BEM is preferable to the use of finite elements in that only the 'boundary' of the problem need be modelled. This results in a dramatic reduction in data preparation costs and reduces possible errors at this important stage. Also the mesh can be modified very easily using BEASY which is important when variations are made in the number, size and position of the holes in the section.

4. Mathematical Solution

4.1 LAMÉ THICK CYLINDER THEORY

Two-dimensional problems in elastostatics arise naturally in two distinct ways. In the first case the body being deformed is a long right cylinder acted upon by external forces which are so arranged that the component of the displacement in the direction of the axis of the cylinder vanishes, and the remaining components remain constant along the length of the cylinder – such a body is said to be in a state of *plane strain*. In the second case the body being deformed is a thin plate acted upon by external forces so distributed that the normal component of stress across the plate vanishes – the plate is said to be in a state of *plane stress*. Our problems fall into the former class, i.e., *plane strain*.

For our plane strain problems it is advantageous to use polar coordinates r, θ. In these coordinates the equations of equilibrium reduce to (see Timoshenko and Goodier [5])

$$\frac{\partial \sigma_r}{\partial r} + \frac{1}{r}\frac{\partial \tau_{r\theta}}{\partial \theta} + \frac{\sigma_r - \sigma_\theta}{r} = 0 \qquad (A4.2)$$

$$\frac{1}{r}\frac{\partial \sigma_r}{\partial \theta} + \frac{\partial \tau_{r\theta}}{\partial r} + \frac{2\tau_{r\theta}}{r} = 0 . \qquad (A4.3)$$

When the body force is zero these equations (A4.2) and (A4.3) are satisfied by putting

$$\sigma_r = \frac{1}{r}\frac{\partial \chi}{\partial r} + \frac{1}{r^2}\frac{\partial^2 \chi}{\partial \theta^2},$$

$$\sigma_\theta = \frac{\partial^2 \chi}{\partial r^2}, \qquad\qquad (A4.4)$$

$$\tau_{r\theta} = \frac{1}{r^2}\frac{\partial \chi}{\partial \theta} - \frac{1}{r}\frac{\partial^2 \chi}{\partial r \partial \theta}.$$

These are valid solutions for plane strain or plane stress if χ, considered as a function of r and θ, satisfies the bi-harmonic equation

$$\nabla_1^4 \chi = 0, \qquad\qquad (A4.5)$$

where ∇_1^2 denotes the two-dimensional Laplacian operator

$$\frac{\partial^2}{\partial x^2} + \frac{\partial^2}{\partial y^2}.$$

In polar coordinates, this becomes

$$\left(\frac{\partial^2}{\partial r^2} + \frac{1}{r}\frac{\partial}{\partial r} + \frac{1}{r^2}\frac{\partial^2}{\partial \theta^2}\right)\left(\frac{\partial^2 \chi}{\partial r^2} + \frac{1}{r}\frac{\partial \chi}{\partial r} + \frac{1}{r^2}\frac{\partial^2 \chi}{\partial \theta^2}\right) = 0. \qquad (A4.6)$$

Now consider the case of stress distribution symmetrical about an axis $r = 0$. Since the distribution is now independent of θ, equation (A4.6) reduces to the ordinary differential equation

$$\frac{d^4 \chi}{dr^4} + \frac{2}{r}\frac{d^3 \chi}{dr^3} - \frac{1}{r^2}\frac{d^2 \chi}{dr^2} + \frac{1}{r^2}\frac{d\chi}{dr} = 0, \qquad (A4.7)$$

which has the general solution

$$\chi = A\ln r + Br^2\ln r + Cr^2 + D. \qquad\qquad (A4.8)$$

Substitution of this function in the expression (A4.4) gives, for axial symmetry, the stress components

$$\sigma_r = A/r^2 + B(1 + 2\ln r) + 2C,$$
$$\sigma_\theta = -A/r^2 + B(3 + 2\ln r) + 2C, \qquad (A4.9)$$

$$\tau_{r\theta} = 0.$$

These results (A4.9) can readily be used to solve the problem of a hollow cylinder subjected to uniform pressures on the inner and outer surfaces. The boundary conditions are

$$\sigma_r = -p_a \text{ at } r = a,$$

$$\sigma_r = -p_b \text{ at } r = b, \tag{A4.10}$$

where $a < b$.

If u, v are the radial and tangential components of the displacement, respectively, then for plane strain the radial component of strain is given by

$$\frac{\partial u}{\partial r} = \frac{1+v}{E} \left[(1-v)\sigma_r - v\sigma_\theta \right]$$
$$= \frac{1+v}{E} \left[A/r^2 + B\left[(1-4v) - (2-4v)\ln r \right] + C(2-4v) \right],$$

where v is Poisson's ratio and E is Young's modulus. Hence

$$u = \frac{1+v}{E} \left[-A/r - Br + B(2-4v)r \ln r + C(2-4v)r \right] + f(\theta), \tag{A4.11}$$

where $f(\theta)$ is a function of θ only.

The tangential component of strain is

$$\frac{u}{r} + \frac{1}{r}\frac{\partial v}{\partial \theta} = \frac{1+v}{E} \left[-v\sigma_r + (1-v)\sigma_\theta \right]$$
$$= \frac{1+v}{E} \left[-A/r^2 + B\left[(3-4v)\ln r \right] + C(2-4v) \right]$$

and, using equation (A4.11), we find that

$$v = \frac{4B(1-v^2)}{E} r\theta - \int f(\theta)\, d\theta + g(r), \tag{A4.12}$$

where $g(r)$ is a function of r only.

Since

$$\tau_{r\theta} = 0, \; \gamma_{r\theta} = \frac{1}{r}\frac{\partial u}{\partial \theta} + \frac{\partial v}{\partial r} - \frac{v}{r} = 0$$

and by substitution from equations (A4.11) and (A4.12)

$$\frac{1}{r}f'(\theta) + g'(r) = \frac{1}{r}\int f(\theta)\,d\theta + \frac{1}{r}g(r)$$

so that

$$g(r) = Fr, \; f(\theta) = H\sin\theta + K\cos\theta,$$

where F, H and K are constants to be determined. Substituting these functions in equation (4.12), we find that the tangential displacement is

$$v = \frac{4B(1-v^2)r\theta}{E} + Fr + H\cos\theta - K\sin\theta \cdot$$

Clearly, the displacement is not single valued unless $B = 0$, and this is the required condition, enabling the stress distribution to be determined unambiguously.

Putting $B = 0$ in the expression (A4.9) and using the boundary conditions (A4.10), we find that

$$A = \frac{(p_b - p_a)a^2b^2}{b^2 - a^2},$$

$$C = \frac{a^2 p_a - b^2 p_b}{2(b^2 - a^2)}, \tag{A4.13}$$

and, by subtraction, the stress components are

$$\sigma_r = \frac{a^2b^2(p_b - p_a)}{(b^2 - a^2)r^2} + \frac{a^2 p_a - b^2 p_b}{b^2 - a^2},$$

$$\sigma_\theta = -\frac{a^2b^2(p_b - p_a)}{(b^2 - a^2)r^2} + \frac{a^2 p_a - b^2 p_b}{b^2 - a^2} \cdot \tag{A4.14}$$

In the present case, the displacement must be axially symmetrical, so that $f(\theta) \equiv 0$. Then, substituting from equations (A4.13) in expressions (A4.11) and (A4.12), the components of displacements are found to be

$$u = \frac{1+v}{E(b^2 - a^2)} \left[\frac{a^2 b^2 (p_a - p_b)}{r} + (1 - 2v)(a^2 p_a - b^2 p_b)r \right], \qquad (A4.15)$$

$$v = Fr, \qquad (A4.16)$$

where the tangential component v is merely a pure rotation.

5. Model Validation

5.1 SQUARE SECTION WITH CENTRAL HOLE

The BEASY analysis uses conditions of plane strain and because of symmetry it is only necessary to consider one quarter of the body. It should be noted that BEASY can also deal with plane stress. The geometry plot with the defined elements is shown in Figure A4.1.

It is important to note that elements are only defined on the actual boundary and not on lines of symmetry. Also quadratic elements with three node points are used. Three elements are placed along the quadrant representing the hole and five along each (half) edge of the block.

A variety of internal points has been defined, namely, two linear arrays, i.e., one horizontal (points 13-22) and one at 45° to the axis (points 23-32) together with one rectangular array in the corner (16 points). The only loading is the $1000 lb / in^2$ pressure on the inside of the hole. (Note that a negative value is used in the data since compression stress is implied).

BEASY has been used to calculate the normal and shear stresses at all internal points together with the corresponding principal stresses. In the latter case it is possible to include values of shear stress calculated according to the von Mises and Tresca theories of shear yielding in complex stress (see Benham and Warnock [1]).

A typical stress plot is given in Figure A4.3 which also shows the variations along the line of internal points 23 to 32 at 45° to the horizontal axis. In order to build up confidence in these results comparisons have been made with solutions to the problem of a pressurized hollow thick cylinder, found by using:

(1) Lamé thick cylinder theory as previously described
(2) thick cylinder analysis in BEASY.

In Lamé thick cylinder theory the stress components σ_r and σ_θ are given by equation (A4.14) and the displacement u is given by equation (A4.15), where

$$a = 1\text{in} , \ b = 5\text{in} , \ p_a = 1000\text{lb/in}^2 , \ p_b = 0$$
$$v = 0.2 , \ E = 50 \times 10^4 \text{lb/in}^2 .$$

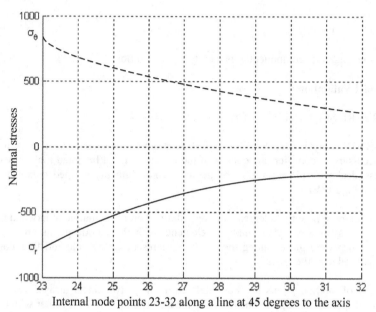

Internal node points 23-32 along a line at 45 degrees to the axis

Figure A4.3. Typical stress plot for square section with hole under internal pressure.

Computed values of normal stresses σ_r and σ_θ at points equivalent to internal points 13 to 22 on the horizontal axis are presented in Table A4.1 by using

(1) BEASY
(2) BEASY thick cylinder analysis
(3) Lamé thick cylinder theory.

Clearly there is good agreement which gives confidence in results at other internal points produced by BEASY. Also computed values of the displacement u at the same internal points produced by BEASY and Lamé thick cylinder theory are presented in Table A4.2 and again there is close agreement.

The actual values of displacements for all mesh points have been calculated using BEASY and Figure A4.4 shows exaggerated displacements indicating the deformed shape and the movement of internal points. As an example, the horizontal displacement at point 7 on element 3, i.e., on the edge of the hole, is computed by BEASY to be 2.5474×10^{-3} in. This compares well with the BEASY thick cylinder analysis value of 2.5677×10^{-3} in and the Lamé thick cylinder theory value of 2.5600×10^{-3} in. The latter value is obtained by using equation (A4.15) with $r = 1$.

Table A4.1. Computed values of normal stresses σ_r and σ_θ at internal points 13 to 22 (see Figure A4.1) using BEASY and Lamé thick cylinder theory.

-	Position	BEASY		BEASY Thick Cylinder		Lamé Thick Cylinder	
Internal	r	σ_r	σ_θ	σ_r	σ_θ	σ_r	σ_θ
Point	in	lb/in^2	lb/in^2	lb/in^2	lb/in^2	lb/in^2	lb/in^2
13	1.100	−821.2	891.0	−820.3	902.1	−819.2	902.5
14	1.211	−670.0	742.0	−668.9	751.6	−668.5	751.8
15	1.322	−555.4	628.4	−554.4	637.3	−554.2	637.5
16	1.433	−466.3	539.9	−465.5	548.6	−465.4	548.7
17	1.544	−395.7	469.8	−395.1	478.3	−395.0	478.4
18	1.656	−338.7	413.2	−338.4	421.6	−338.4	421.7
19	1.767	−292.1	366.9	−292.1	375.3	−292.1	375.4
20	1.878	−253.4	328.5	−253.8	337.0	−253.7	337.1
21	1.989	−221.0	296.4	−221.7	304.9	−221.7	305.0
22	2.100	−193.5	269.2	−194.5	277.8	−194.5	277.9

Table A4.2. Computed values of the displacement at internal points 13 to 22 (see Figure A4.1) using BEASY and Lamé thick cylinder theory.

	Position	BEASY	Lamé Thick Cylinder
Internal	r	$u \, (\times 10^{-3})$	$u \, (\times 10^{-3})$
Point	in	in	in
13	1.100	2.3254	2.3387
14	1.211	2.1238	2.1369
15	1.322	1.9572	1.9701
16	1.433	1.8175	1.8302
17	1.544	1.6990	1.7114
18	1.656	1.5974	1.6094
19	1.767	1.5095	1.5211
20	1.878	1.4329	1.4440
21	1.989	1.3657	1.3763
22	2.100	1.3066	1.3165

5.2 SQUARE SECTION WITH NINE HOLES

This is a generalization of the single hole problem in which nine holes are placed symmetrically in the square section. Again, plane strain is assumed and symmetry means that only one quarter of the section need be considered. The corresponding geometry plot with mesh points is shown in Figure A4.2.

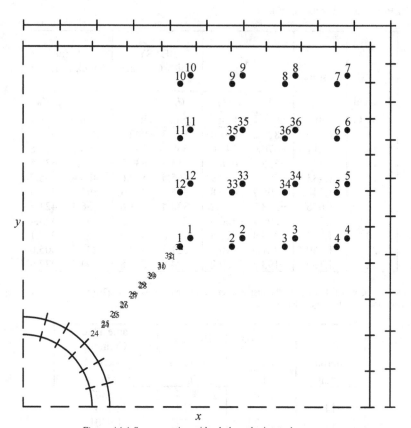

Figure A4.4. Square section with a hole under internal pressure.

It is important to note that the command file was produced by simply editing that for the single hole case. In this way any distribution of holes, whether uniform or random, can readily be described. For the chosen example of nine holes a significant number of internal mesh points have been defined.

A complete set of results as for the single hole case can be obtained including
- normal and shear stresses at internal points;

- principal stresses;
- displacements at boundary;
- displacements internally;
- stress plot for internal points 36 to 45;
- deformed shape.

The stress plot for internal points 36 to 45 is presented in Figure A4.5 and the deformed shape in Figure A4.6.

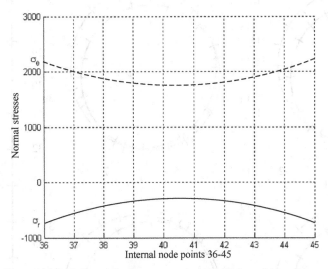

Figure A4.5. Typical stress plot for square section with 9 holes under internal pressure.

6. Interpretation and Conclusions

Using the BEASY software package has meant that the definition of the computer model is a fairly straightforward task. This is an inherent advantage of boundary element analysis over the more usual finite element method. Since elements are defined on the boundary of the plate and holes only, then changing the number of elements or the geometry of the boundaries has little impact on the users data file (.GAT). This contrasts strongly with the FE approach which would imply a totally new mesh to be defined in either of these cases. Similar advantages would be exhibited for three dimensional problems.

It is also worth noting that BEASY permits free choice of the positions of internal points which are not linked to any meshes as is the case for FE analysis. However, they should not be placed too close to the boundary as unpredictable results can sometimes be produced. To enable contouring of stress to be performed it is necessary to have a significant number of internal points evenly spread throughout the geometry.

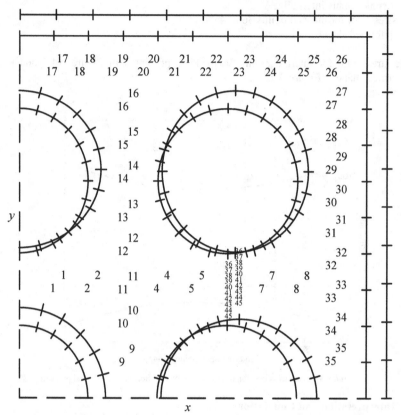

Figure A4.6. Exaggerated displacements indicating the deformed shape and movement of internal points for a square section with 9 holes under internal pressure.

The results obtained for the single hole case based on linear plane strain analysis are excellent. Comparisons with thick cylinder analysis show good agreement on stresses and displacements. This has been backed up by a separate exercise using FE software which is not described in this case study. The free choice of position of internal points in BEASY shows an advantage when a fine description of stress variation is required. The stress gradient near the holes is steep and it is clearly advantageous to be able to define the stress in such regions by having internal points close together.

The square section with nine holes problem has shown the ease with which changes in the model can be made in BEASY. The basic editing of the user command file

(.GAT) required only a few minutes of effort to modify that for the single hole case. An excellent description of the stresses and displacements for the nine hole problem is obtained by using a substantial number of freely defined internal points. In particular, the close definition of stress variation between two of the holes is clearly demonstrated.

The real significance of this work is that it demonstrates the powerfulness of the boundary element software (BEASY) in the computation of stresses and displacements of plates/sections with orderly or randomly arranged holes/cavities with pressure/shear. Problems of this type arise in the food manufacturing industry.

7. Computer Algorithms

The Boundary Element Analysis System (BEASY) software package available from Computational Mechanics Ltd., Southampton, UK was used to carry out the analysis.

8. References and Bibliography

1. Benham, P.P. and Warnock, V., *Mechanics of Solids and Structures*, 2nd ed., Pitman, London, 1976.
2. Brebbia, C.A., *The Boundary Element Method for Engineers*, 2nd ed., Pentech Press, London, 1984.
3. Caldwell, J., "Benchmark for a plate with circular holes," Boundary Elements Communications, **5**, 1994, 227-229.
4. George, J., "Boundary methods back in favour," Engineering Computers, Jan 1985.
5. Timoshenko, S. and Goodier, J.N., *Theory of Elasticity*, McGraw-Hill, London, 1951.

Project B4

MOTION OF FLUID LAYERS

SUMMARY: This project investigates the flow of two parallel layers of oil and water which are located between two plates. Starting from the rest position the top plate is moved at a constant velocity which immediately affects the motion of the neighbouring oil layer. Eventually this will affect the motion of the lower water layer. A mathematical model is formulated which involves parabolic PDEs. Of course, the equations cannot be solved separately in the two layers because of the boundary conditions at the oil-water interface. A numerical solution is obtained using an implicit finite divided difference scheme in terms of time and space. In this way the velocity of the two fluid layers is obtained at various distances from the plates at different times. Clearly the effects of motion will be more noticeable as time proceeds. The results are validated against the steady state solution for large time.

1. Background

Many practical engineering problems particularly in the area of petroleum engineering involve fluid flow. For those problems where two layers are involved, e.g., oil and water, it may be possible to examine the motion by considering fluid layers. An interesting problem arises when a layer of oil and one of water are sandwiched between two parallel horizontal plates. When the top plate is moved at a constant velocity it is possible to predict the motion of both fluid layers at various distances from the plates for different time values.

The equations governing the motion of the fluids will involve parabolic PDE's and it is possible to formulate relationships which hold true at the oil-water interface involving both the velocities and viscosities of the oil and water. This system of equations cannot be solved analytically in the general case because of the boundary conditions at the oil-water interface. However, the PDEs can be solved numerically by a range of finite difference methods.

2. Problem Statement

Consider two parallel horizontal plates spaced 10 cm apart, as shown in Figure B4.1. Initially, both plates and the two fluid layers (oil and water) are at rest. Starting from $t = 0$, the top plate is moved at a constant velocity of 7cm/s. The equations governing the motion of the fluids are

$$\frac{\partial v_{oil}}{\partial t} = \mu_{oil} \frac{\partial^2 v_{oil}}{\partial x^2}$$

and

$$\frac{\partial v_{water}}{\partial t} = \mu_{water} \frac{\partial^2 v_{water}}{\partial x^2}.$$

The following relationships hold true at the oil-water interface

$$v_{oil} = v_{water}$$

and

$$\mu_{oil} \frac{\partial v_{oil}}{\partial x} = \mu_{water} \frac{\partial v_{water}}{\partial x}.$$

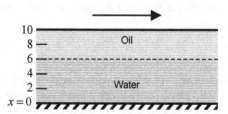

Figure B4.1. Two plates spaced 10 cm apart containing layers of oil and water.

Find the velocity of the two fluid layers at $t = 0.5$, 1 and 1.5s at distances $x = 2$, 4, 6, and 8 cm from the bottom plate. Note that the viscosities μ_{water} and $\mu_{oil} = 1$ and 3 cp, respectively. Validate the numerical results for large time t.

3. Model Formulation

First, we introduce the following key parameters:

$v_{\text{water}}(x,t)$ = velocity of water

$v_{\text{oil}}(x,t)$ = velocity of oil

μ_{water} = viscosity of water

μ_{oil} = viscosity of oil.

We then have two parabolic PDE Initial Boundary Value problems involving the velocities of the two fluids, namely:

$$\begin{cases} \dfrac{\partial v_{\text{water}}}{\partial t} = \mu_{\text{water}} \dfrac{\partial^2 v_{\text{water}}}{\partial x^2}, \ 0 \le x \le 6, \\[2mm] v_{\text{water}}(x,0) = 0, \\[2mm] v_{\text{water}}(0,t) = 0, \\[2mm] v_{\text{water}}(6,t) = v_{\text{oil}}(6,t), \\[2mm] \mu_{\text{oil}} \dfrac{\partial v_{\text{oil}}}{\partial x}\bigg|_{x=6} = \mu_{\text{water}} \dfrac{\partial v_{\text{water}}}{\partial x}\bigg|_{x=6} \end{cases} \tag{B4.1}$$

$$\begin{cases} \dfrac{\partial v_{\text{oil}}}{\partial t} = \mu_{\text{oil}} \dfrac{\partial^2 v_{\text{oil}}}{\partial x^2}, \ 6 \le x \le 10, \\[2mm] v_{\text{oil}}(x,0) = 0, \\[2mm] v_{\text{oil}}(10,t) = 7, \\[2mm] v_{\text{oil}}(6,t) = v_{\text{water}}(6,t), \\[2mm] \mu_{\text{oil}} \dfrac{\partial v_{\text{oil}}}{\partial x}\bigg|_{x=6} = \mu_{\text{water}} \dfrac{\partial v_{\text{water}}}{\partial x}\bigg|_{x=6} \end{cases} \tag{B4.2}$$

We cannot solve the equations (B4.1) and (B4.2) separately, because they share the same boundary conditions at the oil-water interface (i.e., $x = 6$).

The first interface condition $v_{\text{water}}(6,t) = v_{\text{oil}}(6,t)$ confirms the positional continuity of velocity of the two fluids at the interface. The second interface condition $\mu_{\text{water}} \dfrac{\partial v_{\text{water}}}{\partial x}\big|_{x=6} = \mu_{\text{oil}} \dfrac{\partial v_{\text{oil}}}{\partial x}\big|_{x=6}$ refers to the tangent at the oil-water interface. If μ_{water} and μ_{oil} are chosen to be the same value (i.e., the same fluid), this confirms the tangential continuity of velocity of two fluids at the interface. With $\mu_{\text{water}} \ne \mu_{\text{oil}}$, we have

$$\frac{\partial v_{\text{water}}}{\partial x}\Big|_{x=6} = \frac{\mu_{\text{oil}}}{\mu_{\text{water}}} \frac{\partial v_{\text{oil}}}{\partial x}\Big|_{x=6}, \tag{B4.3}$$

and so the tangents differ by a coefficient ratio of $\mu_{\text{oil}} / \mu_{\text{water}}$ at the interface of the fluids.

To formulate a mathematical model we make the following assumptions:

A1. No viscosity at the top plate.
A2. The top plate is moved at a constant velocity of 7cm/s throughout.
A3. At the oil-water interface (i.e., $v_{\text{water}\,m}^{l+1}$, $v_{\text{oil}\,m}^{l+1}$), this process introduces nodes that lie outside the interface, i.e., $v_{\text{water}\,m+1}^{l+1}$ and $v_{\text{oil}\,m-1}^{l+1}$ are introduced.

To obtain the solution numerically, we apply an implicit scheme on equations (B4.1) and (B4.2) with finite divided differences in time and space, respectively. This gives

$$\frac{v_{\text{water}\,i}^{l+1} - v_{\text{water}\,i}^{l}}{\Delta t} = \mu_{\text{water}} \frac{v_{\text{water}\,i+1}^{l+1} - 2v_{\text{water}\,i}^{l+1} + v_{\text{water}\,i-1}^{l+1}}{(\Delta x)^2}, \; i = 1, 2, \cdots, m, \tag{B4.4}$$

$$\frac{v_{\text{oil}\,j}^{l+1} - v_{\text{oil}\,j}^{l}}{\Delta t} = \mu_{\text{oil}} \frac{v_{\text{oil}\,j+1}^{l+1} - 2v_{\text{oil}\,j}^{l+1} + v_{\text{oil}\,j-1}^{l+1}}{(\Delta x)^2}, \; j = m, m+1, \cdots, n-1, \tag{B4.5}$$

which can be expressed as

$$\left(\frac{\mu_{\text{water}}}{(\Delta x)^2}\right)v_{\text{water}\,i-1}^{l+1} - \left(\frac{2\mu_{\text{water}}}{(\Delta x)^2} + \frac{1}{\Delta t}\right)v_{\text{water}\,i}^{l+1} + \left(\frac{\mu_{\text{water}}}{(\Delta x)^2}\right)v_{\text{water}\,i+1}^{l+1} = -\frac{1}{\Delta t}v_{\text{water}\,i}^{l}, \tag{B4.6}$$

for $i = 1, 2, \cdots, m$,

$$\left(\frac{\mu_{\text{oil}}}{(\Delta x)^2}\right)v_{\text{oil}\,j-1}^{l+1} - \left(\frac{2\mu_{\text{oil}}}{(\Delta x)^2} + \frac{1}{\Delta t}\right)v_{\text{oil}\,j}^{l+1} + \left(\frac{\mu_{\text{oil}}}{(\Delta x)^2}\right)v_{\text{oil}\,j+1}^{l+1} = -\frac{1}{\Delta t}v_{\text{oil}\,j}^{l}, \tag{B4.7}$$

for $j = m, m+1, \cdots, n-1$, respectively.

Starting from the bottom of the water, i.e., $i = 1$,

$$-\left(\frac{2\mu_{\text{water}}}{(\Delta x)^2} + \frac{1}{\Delta t}\right)v_{\text{water}\,1}^{l+1} + \left(\frac{\mu_{\text{water}}}{(\Delta x)^2}\right)v_{\text{water}\,2}^{l+1} = -\frac{1}{\Delta t}v_{\text{water}\,1}^{l} - \left(\frac{\mu_{\text{water}}}{(\Delta x)^2}\right)v_{\text{water}\,0}^{l+1}. \tag{B4.8}$$

From the first boundary condition of equation (B4.1), we have $v_{\text{water}}(0,t) = v_{\text{water}\,0}^{l+1} = 0$.

Therefore, equation (B4.8) becomes

$$-\left(\frac{2\mu_{\text{water}}}{(\Delta x)^2} + \frac{1}{\Delta t}\right)v_{\text{water}\,1}^{l+1} + \left(\frac{\mu_{\text{water}}}{(\Delta x)^2}\right)v_{\text{water}\,2}^{l+1} = -\frac{1}{\Delta t}v_{\text{water}\,1}^{l}. \tag{B4.9}$$

A similar exercise can be performed for the interface, where the original difference equations are

$$\left(\frac{\mu_{\text{water}}}{(\Delta x)^2}\right)v_{\text{water}\,m-1}^{l+1} - \left(\frac{2\mu_{\text{water}}}{(\Delta x)^2} + \frac{1}{\Delta t}\right)v_{\text{water}\,m}^{l+1} + \left(\frac{\mu_{\text{water}}}{(\Delta x)^2}\right)v_{\text{water}\,m+1}^{l+1} = -\frac{1}{\Delta t}v_{\text{water}\,m}^{l}, \tag{B4.10}$$

$$\left(\frac{\mu_{\text{oil}}}{(\Delta x)^2}\right)v_{\text{oil}\,m-1}^{l+1} - \left(\frac{2\mu_{\text{oil}}}{(\Delta x)^2} + \frac{1}{\Delta t}\right)v_{\text{oil}\,m}^{l+1} + \left(\frac{\mu_{\text{oil}}}{(\Delta x)^2}\right)v_{\text{oil}\,m+1}^{l+1} = -\frac{1}{\Delta t}v_{\text{oil}\,m}^{l}. \tag{B4.11}$$

The $v_{\text{water}\,m+1}^{l+1}$ and $v_{\text{oil}\,m-1}^{l+1}$ can be removed by invoking the second interface condition. At the oil-water interface, the following condition must hold

$$\mu_{\text{water}}\frac{\partial v_{\text{water}}}{\partial x}\bigg|_{x=6} = \mu_{\text{oil}}\frac{\partial v_{\text{oil}}}{\partial x}\bigg|_{x=6}. \tag{B4.12}$$

A finite divided difference can be substituted for the derivatives to give

$$\mu_{\text{water}}\frac{v_{\text{water}\,m+1}^{l+1} - v_{\text{water}\,m-1}^{l+1}}{2\Delta x} = \mu_{\text{oil}}\frac{v_{\text{oil}\,m+1}^{l+1} - v_{\text{oil}\,m-1}^{l+1}}{2\Delta x}, \tag{B4.13}$$

which can be solved to yield

$$v_{\text{water}\,m+1}^{l+1} = \frac{\mu_{\text{oil}}}{\mu_{\text{water}}}(v_{\text{oil}\,m+1}^{l+1} - v_{\text{oil}\,m-1}^{l+1}) + v_{\text{water}\,m-1}^{l+1}. \tag{B4.14}$$

Substituting equation (B4.14) into equation (B4.10), gives

$$\left(\frac{2\mu_{\text{water}}}{(\Delta x)^2}\right)v_{\text{water}\,m-1}^{l+1} - \left(\frac{2\mu_{\text{water}}}{(\Delta x)^2} + \frac{1}{\Delta t}\right)v_{\text{water}\,m}^{l+1} + \left(\frac{\mu_{\text{oil}}}{(\Delta x)^2}\right)(v_{\text{oil}\,m+1}^{l+1} - v_{\text{oil}\,m-1}^{l+1})$$

$$= -\frac{1}{\Delta t}v_{\text{water}\,m}^{l}. \tag{B4.15}$$

Also, the first interface condition tells us

$$v_{\text{water}\,m}^{l+1} = v_{\text{oil}\,m}^{l+1}. \tag{B4.16}$$

Adding equations (B6.11) and (B.15) together, yields

$$\left(\frac{2\mu_{\text{water}}}{(\Delta x)^2}\right) v_{\text{water}\,m-1}^{l+1} - \left(\frac{2\mu_{\text{water}} + 2\mu_{\text{oil}}}{(\Delta x)^2} + \frac{2}{\Delta t}\right) v_{\text{water}\,m}^{l+1} + \left(\frac{2\mu_{\text{oil}}}{(\Delta x)^2}\right) v_{\text{oil}\,m+1}^{l+1}$$

$$= -\frac{1}{\Delta t} v_{\text{water}\,m}^{l} - \frac{1}{\Delta t} v_{\text{oil}\,m}^{l}. \tag{B4.17}$$

Starting from the oil-water interface, i.e., $j = m+1$,

$$\left(\frac{\mu_{\text{oil}}}{(\Delta x)^2}\right) v_{\text{oil}\,m}^{l+1} - \left(\frac{2\mu_{\text{oil}}}{(\Delta x)^2} + \frac{1}{\Delta t}\right) v_{\text{oil}\,m+1}^{l+1} + \left(\frac{\mu_{\text{oil}}}{(\Delta x)^2}\right) v_{\text{oil}\,m+2}^{l+1}$$

$$= -\frac{1}{\Delta t} v_{\text{oil}\,m+1}^{l}. \tag{B4.18}$$

A similar exercise can be performed for the top plate, where the original difference equation is

$$\left(\frac{\mu_{\text{oil}}}{(\Delta x)^2}\right) v_{\text{oil}\,n-2}^{l+1} - \left(\frac{2\mu_{\text{oil}}}{(\Delta x)^2} + \frac{1}{\Delta t}\right) v_{\text{oil}\,n-1}^{l+1} = -\frac{1}{\Delta t} v_{\text{oil}\,m+1}^{l} - \left(\frac{\mu_{\text{oil}}}{(\Delta x)^2}\right) v_{\text{oil}\,n}^{l+1}. \tag{B4.19}$$

From the first boundary condition of equation (B4.2), we have $v_{\text{oil}}(10,t) = v_{\text{oil}\,n}^{l+1} = 7$.

Therefore, equation (B4.19) becomes

$$\left(\frac{\mu_{\text{oil}}}{(\Delta x)^2}\right) v_{\text{oil}\,n-2}^{l+1} - \left(\frac{2\mu_{\text{oil}}}{(\Delta x)^2} + \frac{1}{\Delta t}\right) v_{\text{oil}\,n-1}^{l+1} = -\frac{1}{\Delta t} v_{\text{oil}\,m+1}^{l} - \left(\frac{7\mu_{\text{oil}}}{(\Delta x)^2}\right). \tag{B4.20}$$

The initial conditions of equations (B4.1) and (B4.2) tell us that $v_{\text{water}\,i}^{l}$ and $v_{\text{oil}\,j}^{l} = 0$ for all $i = 1, 2, \cdots, m$, $j = m, m+1, \cdots, n-1$.

When equations (B4.6), (B4.7), (B4.9), (B4.17), (B4.18) and (B4.20) are written for all the interior nodes, we have a resulting system of $(n-1)$ equations in $(n-1)$ unknowns. In addition, the system is tridiagonal. Thus, we can apply the Thomas algorithm for tridiagonal systems.

Table B4.1 shows the velocity of the two fluid layers at $t = 0.5$, 1 and 1.5s for distances $x = 2$, 4, 6, and 8 cm from the bottom plate with μ_{water} and $\mu_{oil} = 1$ and 3 cp, respectively.

Table B4.1 Velocity at distance x = 2, 4, 6, 8cm for different times.

Velocity cm/s	$x = 2$cm	$x = 4$cm	$x = 6$cm	$x = 8$cm
$t = 0.5$s	0.0012	0.0200	0.3386	1.3809
$t = 1.0$s	0.0158	0.1189	0.8792	2.2588
$t = 1.5$s	0.0491	0.2611	1.3384	2.8202

Figure B4.2 shows the velocity of the two fluid layers at $t = 0.5$, 5 and 50s for distances $0 \le x \le 10$, with μ_{water} and $\mu_{oil} = 1$ and 3 cp, respectively.

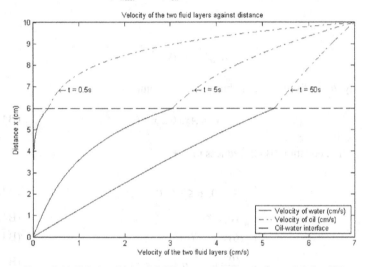

Figure B4.2. Velocity of the two fluid layers against distance for t = 0.5, 5 and 50s.

As expected, the velocity increases as the time increases. If t is allowed to be sufficiently large, then we approach the steady state situation and equations (B4.1) and (B4.2) reduce to simple ODE's and their solutions are given by

$$v_{water}(x) = \frac{7\mu_{oil}}{6\mu_{oil} + 4\mu_{water}} x, \ 0 \le x \le 6, \tag{B4.21}$$

$$v_{oil}(x) = \frac{7\mu_{water}}{6\mu_{oil} + 4\mu_{water}} x + \frac{21(\mu_{oil} - \mu_{water})}{3\mu_{oil} + 2\mu_{water}}, \ 6 \le x \le 10. \tag{B4.22}$$

The derivation of these equations is given in Section 5.

5. Model Validation

In this project the two systems of PDEs (B4.1) and (B4.2) cannot be solved analytically but for model validation purposes we can compare the numerical solution in Section 4 against the steady state solution for large time. First, we should derive the steady state solution given by equations (B4.21) and (B4.22).

In the steady state case, the set of equations (B4.1) reduces to

$$\frac{d^2 v_{water}}{dx^2} = 0, \ 0 \leq x \leq 6, \tag{B4.23}$$

$$v_{water}(0) = 0. \tag{B4.24}$$

Hence

$$v_{water}(x) = Ax + B$$

and clearly $B = 0$ from (B4.24), thus giving the solution

$$v_{water}(x) = Ax, \ 0 \leq x \leq 6. \tag{B4.25}$$

Now the set of equation (B4.2) reduces to

$$\frac{d^2 v_{oil}}{dx^2} = 0, \ 6 \leq x \leq 10, \tag{B4.26}$$

$$v_{oil}(10) = 7, \tag{B4.27}$$

$$v_{oil}(6) = v_{water}(6) \tag{B4.28}$$

$$\mu_{oil} \frac{dv_{oil}}{dx}\bigg|_{x=6} = \mu_{water} \frac{dv_{water}}{dx}\bigg|_{x=6}. \tag{B4.29}$$

Hence

$$v_{oil}(x) = Cx + D, \ 6 \leq x \leq 10 \tag{B4.30}$$

and the boundary conditions (B4.27) and (B4.28) give

$$10C + D = 7 \tag{B4.31}$$

and

$$6C + D = 6A. \tag{B4.32}$$

Finally, equation (B4.29) gives

$$\mu_{oil}(C) = \mu_{water}(A).$$ (B4.33)

On solving (B4.31) – (B4.33), we obtain

$$A = \frac{7\mu_{oil}}{6\mu_{oil} + 4\mu_{water}}$$

$$C = \frac{7\mu_{water}}{6\mu_{oil} + 4\mu_{water}}$$

$$D = \frac{21(\mu_{oil} - \mu_{water})}{3\mu_{oil} + 2\mu_{water}}.$$

So the final steady state solution is

$$v_{water}(x) = \frac{7\mu_{oil}}{6\mu_{oil} + 4\mu_{water}} x, \; 0 \le x \le 6,$$ (B4.34)

$$v_{oil}(x) = \frac{7\mu_{water}}{6\mu_{oil} + 4\mu_{water}} x + \frac{21(\mu_{oil} - \mu_{water})}{3\mu_{oil} + 2\mu_{water}}, \; 6 \le x \le 10.$$ (B4.35)

Now we produce computer graphs of the numerical results from Section 4 for large time t. Clearly, they approach the above linear steady state results (see Figure B4.3).

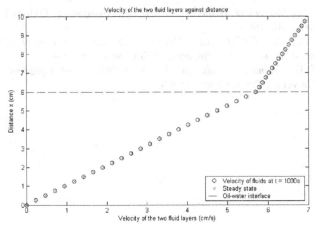

Figure B4.3. Large time behaviour of velocity of two fluid layers against distance for t = 1000s.

6. Interpretation and Conclusions

This project demonstrates how it is possible to deal with the motion of two adjacent fluid layers (in this case, oil and water) in cases where we have prescribed knowledge of the velocities and viscosities of the two fluids involved. Inevitably a mathematical model will result in the problem being formulated in terms of PDEs with prescribed boundary and initial conditions provided we make a number of simplifying assumptions. A finite difference solution will result in a system of linear equations, which, in this case, is tridiagonal. For efficiency, the Thomas algorithm is then called into play. For model validation purposes the results can be extended and compared to the steady state solution for large values of time. Clearly, this project should help to promote ideas on how to deal with other more realistic situations involving the motion of fluid layers.

7. Computer Algorithms

A computer program was developed for the finite difference method for the solution of the PDEs. Again, this resulted in the solution of a tridiagonal system of linear equations using the Thomas algorithm (see Section 7 of Project B1).

8. References and Bibliography

1. Albertson, M.L., Barton, J.R. and Simons, D.B., *Fluid Mechanics for Engineers*, Prentice-Hall, Englewood Cliffs, N.J., 1960.
2. Cussler, E.L., *Diffusion: Mass Transfer in Fluid Systems*, 2nd ed., Cambridge University Press, New York, 1997.
3. Jaluria, Y., *Computer Methods for Engineering*, Allyn and Bacon, Inc., Boston, 1988.
4. Köckler, N., *Numerical Methods and Scientific Computing*, Oxford University Press, New York, 1994.
5. Lapidus, L. and Pinder, G.F., *Numerical Solution of Partial Differential Equations in Science and Engineering*, Wiley, New York, 1981.
6. Smith, G.D., *Numerical Solution of Partial Differential Equation*, 2nd ed., Oxford University Press, London, 1969.

Project B5

MASS BALANCE OF A REACTOR WITH TIME DEPENDENCY

SUMMARY: Project B1 considered the one-dimensional mass balance of a cylindrical chemical reactor in steady state. This meant that the governing parabolic PDE reduced to an ODE. This project now extends the work in Project B1 to include time dependency which involves the full PDE. An analytical solution is no longer possible but a numerical solution is obtained using finite difference methods. As a result computer graphs are plotted which show the variation of concentration of the chemical with distance along the longitudinal axis of the reactor at different times. For model validation purposes, these results for large time are compared with the exact results obtained in Project B1 for the steady state case.

1. Background

The background to this project is the same as that for Project B1. The previous work is now extended to deal with a chemical whose concentration depends on both distance and time. This means that we have to deal with the full parabolic PDE and not the simplified steady state case of the ODE. Figure B1.1 in Project B1 still applies.

2. Problem Statement

Figure B1.1 in Project B1 shows an elongated cylindrical reactor with a single entry and exit point. Using the same modelling assumptions as in Project B1, the same mass balance can be performed on a finite segment of length Δx to produce equation (B1.1).

Use this mass balance in the limit as Δx and Δt tend to zero to formulate the governing parabolic PDE and the associated boundary and initial conditions. State clearly any modelling assumptions made in the formulation. Then consider the solution of this PDE.

Use finite differences to produce a numerical solution of this parabolic PDE initial-boundary value problem and hence show how the concentration of chemical varies with distance at different times during the build-up of chemical in the reactor.

Finally, validate the model with reference to the steady state solution obtained in Project B1.

3. Model Formulation

To formulate a mathematical model we make the following assumptions:

A1. Chemical being modelled is subject to first-order decay.
A2. The tank is well mixed vertically and laterally.
A3. Dispersion in the reactor does not affect the exit rate.
A4. Prior to $t = 0$ the reactor is filled with water which contains no chemical.
A5. Starting from $t = 0$ the chemical is injected into the reactor's inflow at a constant level of c_{in}.

Then as $\Delta x \to 0$ and $\Delta t \to 0$, and using the modelling assumptions A1 and A2, equation (B1.1) approaches in the limit

$$\frac{\partial c}{\partial t} = D\frac{\partial^2 c}{\partial x^2} - U\frac{\partial c}{\partial x} - \gamma c, \ 0 < x < L, \tag{B5.1}$$

where $U = F / A_c$ is the velocity of the water flowing through the tank.

It should be noted here that the steady state case of this equation (B5.1) has been solved in Project B1.

The modelling assumption A3 means that chemical leaves the reactor purely as a function of flow through the outlet pipe. This leads to the boundary condition

$$c'(L,t) = 0, \ t > 0. \tag{B5.2}$$

The modelling assumption A4 leads to the initial condition

$$c(x,0) = 0, \ 0 < x < L. \tag{B5.3}$$

Finally, the modelling assumption A5 leads to the boundary condition

$$c(0,t) = c_{in} + \frac{D}{U}c'(0,t), \ t > 0. \tag{B5.4}$$

Hence we have the initial-boundary value problem

$$\begin{cases} \dfrac{\partial c}{\partial t} = D\dfrac{\partial^2 c}{\partial x^2} - U\dfrac{\partial c}{\partial x} - \gamma c, \ 0 < x < L, \\ c(x,0) = 0, \ 0 < x < L, \\ c(0,t) = c_{in} + \dfrac{D}{U}c'(0,t), \ t > 0, \\ c'(L,t) = 0, \ t > 0. \end{cases} \tag{B5.5}$$

4. Numerical Solution

The PDE together with initial and boundary conditions given in equation (B5.5) cannot be solved analytically. Just as for the steady state ODE case in Project B1, the parabolic PDE can be solved by substituting finite divided differences for the partial derivatives. However, in contrast to the ODE, we must now consider changes in time as well as in space. Because of their time-variable nature, solutions to these equations involve a number of new issues, notably stability. We present two fundamental solution approaches based on numerical accuracy — **Implicit and Crank-Nicolson schemes**.

4.1 IMPLICIT METHOD

In the implicit method, the spatial derivative is approximated at an advanced time level $l+1$ (see Figure B5.1).

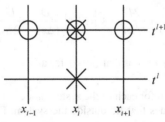

Figure B5.1. A computational molecule for the simple implicit method (×: Grid point involved in time difference, o: Grid point involved in space difference).

A forward finite divided difference is used to approximate the time derivative, namely

$$\frac{\partial c}{\partial t} \cong \frac{c_i^{l+1} - c_i^l}{\Delta t}, \tag{B5.6}$$

which has an error of $O(\Delta t)$.

The first and second space derivatives would be approximated by

$$\frac{\partial c}{\partial x} \cong \frac{c_{i+1}^{l+1} - c_{i-1}^{l+1}}{2\Delta x}, \tag{B5.7}$$

$$\frac{\partial^2 c}{\partial x^2} \cong \frac{c_{i+1}^{l+1} - 2c_i^{l+1} + c_{i-1}^{l+1}}{(\Delta x)^2}, \tag{B5.8}$$

which are second-order accurate. When these approximations are substituted into the original PDE, the resulting difference equation contains several unknowns. Thus, it cannot be solved explicitly by simple algebraic rearrangement. Instead, the entire system of equations must be solved simultaneously. This is possible because, along with the boundary conditions, the implicit formulation results in a set of linear algebraic equations with the same number of equations as unknowns. Thus, the method reduces to the solution of a set of simultaneous linear equations at each point in time.

To illustrate how this is done, we substitute equations (B5.6)-(B5.8) into (B5.5) to give

$$\frac{c_i^{l+1} - c_i^l}{\Delta t} = D\frac{c_{i+1}^{l+1} - 2c_i^{l+1} + c_{i-1}^{l+1}}{(\Delta x)^2} - U\frac{c_{i+1}^{l+1} - c_{i-1}^{l+1}}{2\Delta x} - \gamma c_i^{l+1}, \tag{B5.9}$$

which can be expressed as

$$\left(\frac{D}{(\Delta x)^2} + \frac{U}{2\Delta x}\right)c_{i-1}^{l+1} - \left(\frac{1}{\Delta t} + \frac{2D}{(\Delta x)^2} + \gamma\right)c_i^{l+1} + \left(\frac{D}{(\Delta x)^2} - \frac{U}{2\Delta x}\right)c_{i+1}^{l+1} = -\frac{1}{\Delta t}c_i^l. \tag{B5.10}$$

The initial condition $c(x,0)$ tells us that $c_i^l = 0$ for all $i = 0, 1, \cdots, n$; $l = 0$.

This equation can be written for each of the system nodes. At the ends of the reactor, this process introduces nodes that lie outside the system. For example, at the inlet node ($i = 0$),

$$\left(\frac{D}{(\Delta x)^2} + \frac{U}{2\Delta x}\right)c_{-1}^{l+1} - \left(\frac{1}{\Delta t} + \frac{2D}{(\Delta x)^2} + \gamma\right)c_0^{l+1} + \left(\frac{D}{(\Delta x)^2} - \frac{U}{2\Delta x}\right)c_1^{l+1} = -\frac{1}{\Delta t}c_0^l. \tag{B5.11}$$

The c_{-1}^{l+1} can be removed by invoking the boundary condition (B5.4). At the inlet, the following mass balance must hold

$$c(0,t) = c_{in} + \frac{D}{U}c'(0,t), \qquad (B5.12)$$

where $c(0,t)$ = concentration at $x = 0$ and for $t > 0$. Thus, this boundary condition specifies that the amount of chemical carried into the tank by advection through the pipe must be equal to the amount carried away from the inlet by both advection and turbulent dispersion in the tank. A finite divided difference can be substituted for the derivative, namely

$$c_0^{l+1} = c_{in} + \frac{D}{U}\frac{c_1^{l+1} - c_{-1}^{l+1}}{2\Delta x}, \qquad (B5.13)$$

which can be solved to give

$$c_{-1}^{l+1} = c_1^{l+1} - \frac{2U\Delta x}{D}(c_0^{l+1} - c_{in}). \qquad (B5.14)$$

This can be substituted into equation (B5.11) to give

$$-\left(\frac{1}{\Delta t} + \frac{2D}{(\Delta x)^2} + \gamma + \frac{2U}{\Delta x} + \frac{U^2}{D}\right)c_0^{l+1} + \left(\frac{2D}{(\Delta x)^2}\right)c_1^{l+1} = -\left(\frac{2U}{\Delta x} + \frac{U^2}{D}\right)c_{in} - \frac{1}{\Delta t}c_0^l. \quad (B5.15)$$

A similar exercise can be performed for the outlet, where the original difference equation is

$$\left(\frac{D}{(\Delta x)^2} + \frac{U}{2\Delta x}\right)c_{n-1}^{l+1} - \left(\frac{1}{\Delta t} + \frac{2D}{(\Delta x)^2} + \gamma\right)c_n^{l+1} + \left(\frac{D}{(\Delta x)^2} - \frac{U}{2\Delta x}\right)c_{n+1}^{l+1} = -\frac{1}{\Delta t}c_n^l. \quad (B5.16)$$

The boundary condition at the outlet is

$$c'(L,t) = 0. \qquad (B5.17)$$

As with the inlet, a divided difference can be used to approximate the derivative as follows:

$$\frac{c_{n+1}^{l+1} - c_{n-1}^{l+1}}{2\Delta x} = 0. \qquad (B5.18)$$

Inspection of this equation leads us to conclude that $c_{n+1}^{l+1} = c_{n-1}^{l+1}$. In other words, the slope at the outlet must be zero for equation (B5.18) to hold. Substituting this result into equation (B5.16) and simplifying gives

$$\left(\frac{2D}{(\Delta x)^2}\right)c_{n-1}^{l+1} - \left(\frac{1}{\Delta t} + \frac{2D}{(\Delta x)^2} + \gamma\right)c_n^{l+1} = -\frac{1}{\Delta t}c_n^l. \tag{B5.19}$$

Equations (B5.10), (B5.15) and (B5.19) are written for all the interior nodes resulting in a system of n equations in n unknowns. In addition, the method has the added bonus that the system is tridiagonal. Thus, we can apply the Thomas algorithm again for tridiagonal systems.

Whereas the implicit method described is stable and convergent, it has the defect that the time difference approximation is first-order accurate, whereas the spatial difference approximation is second-order accurate. To remedy this situation, we present an alternative implicit method, namely, the Crank-Nicolson method.

4.2 CRANK-NICOLSON METHOD

The *Crank-Nicolson method* provides an alternative implicit scheme that is second-order accurate in both space and time. To provide this accuracy, difference approximations are developed at the midpoint of the time increment (see Figure B5.2).

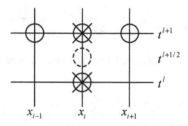

Figure B5.2. A computational molecule for the Crank-Nicolson method (×: Grid point involved in time difference, o: Grid point involved in space difference).

To do this, the first time derivative can be approximated at $t^{l+1/2}$ by

$$\frac{\partial c}{\partial t} \cong \frac{c_i^{l+1} - c_i^l}{\Delta t}. \tag{B5.20}$$

The first and second derivatives in space can be determined at the midpoint by averaging the difference approximations at the beginning (t^l) and at the end (t^{l+1}) of the time increment as follows:

$$\frac{\partial c}{\partial x} \cong \frac{1}{2}\left[\frac{c_{i+1}^{l} - c_{i-1}^{l}}{2\Delta x} + \frac{c_{i+1}^{l+1} - c_{i-1}^{l+1}}{2\Delta x}\right], \tag{B5.21}$$

$$\frac{\partial^2 c}{\partial x^2} \cong \frac{1}{2}\left[\frac{c_{i+1}^{l} - 2c_{i}^{l} + c_{i-1}^{l}}{(\Delta x)^2} + \frac{c_{i+1}^{l+1} - 2c_{i}^{l+1} + c_{i-1}^{l+1}}{(\Delta x)^2}\right]. \tag{B5.22}$$

Substituting equations (B5.20)-(B5.22) into equation (B5.5) gives

$$\frac{c_{i}^{l+1} - c_{i}^{l}}{\Delta t} = \frac{D}{2}\left[\frac{c_{i+1}^{l} - 2c_{i}^{l} + c_{i-1}^{l}}{(\Delta x)^2} + \frac{c_{i+1}^{l+1} - 2c_{i}^{l+1} + c_{i-1}^{l+1}}{(\Delta x)^2}\right]$$
$$-\frac{U}{2}\left[\frac{c_{i+1}^{l} - c_{i-1}^{l}}{2\Delta x} + \frac{c_{i+1}^{l+1} - c_{i-1}^{l+1}}{2\Delta x}\right] - \frac{\gamma}{2}[c_{i}^{l} + c_{i}^{l+1}], \tag{B5.23}$$

which can be expressed as

$$\left(\frac{D}{2(\Delta x)^2} + \frac{U}{4\Delta x}\right)c_{i-1}^{l+1} - \left(\frac{1}{\Delta t} + \frac{D}{(\Delta x)^2} + \frac{\gamma}{2}\right)c_{i}^{l+1} + \left(\frac{D}{2(\Delta x)^2} - \frac{U}{4\Delta x}\right)c_{i+1}^{l+1}$$
$$= -\left(\frac{D}{2(\Delta x)^2} + \frac{U}{4\Delta x}\right)c_{i-1}^{l} - \left(\frac{1}{\Delta t} - \frac{D}{(\Delta x)^2} - \frac{\gamma}{2}\right)c_{i}^{l} - \left(\frac{D}{2(\Delta x)^2} - \frac{U}{4\Delta x}\right)c_{i+1}^{l}. \tag{B5.24}$$

Again, the initial condition $c(x,0)$ tells us that $c_i^l = 0$ for all $i = 0, 1, \cdots, n$; $l = 0$.

At the inlet node ($i = 0$), we have

$$\left(\frac{D}{2(\Delta x)^2} + \frac{U}{4\Delta x}\right)c_{-1}^{l+1} - \left(\frac{1}{\Delta t} + \frac{D}{(\Delta x)^2} + \frac{\gamma}{2}\right)c_{0}^{l+1} + \left(\frac{D}{2(\Delta x)^2} - \frac{U}{4\Delta x}\right)c_{1}^{l+1}$$
$$= -\left(\frac{D}{2(\Delta x)^2} + \frac{U}{4\Delta x}\right)c_{-1}^{l} - \left(\frac{1}{\Delta t} - \frac{D}{(\Delta x)^2} - \frac{\gamma}{2}\right)c_{0}^{l} - \left(\frac{D}{2(\Delta x)^2} - \frac{U}{4\Delta x}\right)c_{1}^{l}. \tag{B5.25}$$

Substituting c_{-1}^{l+1} from equation (B5.14) into equation (B5.25), gives

$$-\left(\frac{1}{\Delta t} + \frac{D}{(\Delta x)^2} + \frac{\gamma}{2} + \frac{U}{\Delta x} + \frac{U^2}{2D}\right)c_{0}^{l+1} + \left(\frac{D}{(\Delta x)^2}\right)c_{1}^{l+1}$$
$$= -\left(\frac{1}{\Delta t} - \frac{D}{(\Delta x)^2} - \frac{\gamma}{2} - \frac{U}{\Delta x} - \frac{U^2}{2D}\right)c_{0}^{l} - \left(\frac{D}{(\Delta x)^2}\right)c_{1}^{l} - 2\left(\frac{D}{\Delta x} + \frac{U}{2}\right)\frac{U}{D}c_{\text{in.}} \tag{B5.26}$$

A similar exercise can be performed for the outlet, where the original difference equation is

$$\left(\frac{D}{2(\Delta x)^2}+\frac{U}{4\Delta x}\right)c_{n-1}^{l+1}-\left(\frac{1}{\Delta t}+\frac{D}{(\Delta x)^2}+\frac{\gamma}{2}\right)c_n^{l+1}+\left(\frac{D}{2(\Delta x)^2}-\frac{U}{4\Delta x}\right)c_{n+1}^{l+1}$$

$$=-\left(\frac{D}{2(\Delta x)^2}+\frac{U}{4\Delta x}\right)c_{n-1}^{l}-\left(\frac{1}{\Delta t}-\frac{D}{(\Delta x)^2}-\frac{\gamma}{2}\right)c_n^{l}-\left(\frac{D}{2(\Delta x)^2}-\frac{U}{4\Delta x}\right)c_{n+1}^{l}. \quad \text{(B5.27)}$$

Substituting $c_{n+1}^{l+1}=c_{n-1}^{l+1}$ into equation (B5.27), gives

$$\left(\frac{D}{(\Delta x)^2}\right)c_{n-1}^{l+1}-\left(\frac{1}{\Delta t}+\frac{D}{(\Delta x)^2}+\frac{\gamma}{2}\right)c_n^{l+1}=-\left(\frac{D}{(\Delta x)^2}\right)c_{n-1}^{l}-\left(\frac{1}{\Delta t}-\frac{D}{(\Delta x)^2}-\frac{\gamma}{2}\right)c_n^{l}. \quad \text{(B5.28)}$$

Although equations (B5.24), (B5.26) and (B5.28) are slightly more complicated than equations (B5.10), (B5.15) and (B5.19), the system is also tridiagonal and, therefore, can be solved efficiently.

5. Model Validation

The numerical results (using implicit and Crank-Nicolson schemes) are plotted in Figure B5.3 for $D=2$, $U=2$, $\gamma=0.2$, $c_{\text{in}}=100$ and $\Delta x=0.25$, where the concentration in the tank is 0 at time zero. As expected, the immediate impact is near the inlet. With increasing time values, the solution eventually approaches the steady state level .

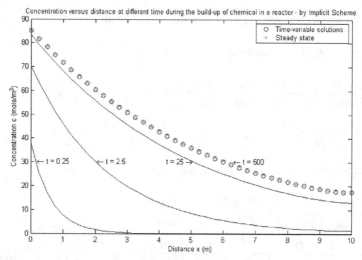

Figure B5.3a. Concentration versus distance at different times during the build-up of chemical in a reactor by implicit scheme.

To assess the accuracy of the finite difference schemes, Table B5.1 presents the results to 14D at $t = 500$ from the implicit and Crank-Nicolson methods both using intervals $\Delta x = 0.25$, $\Delta t = 0.025$, for the case $D = 1$, $U = 1$, $\gamma = 0.2$, $c_{in} = 100$.

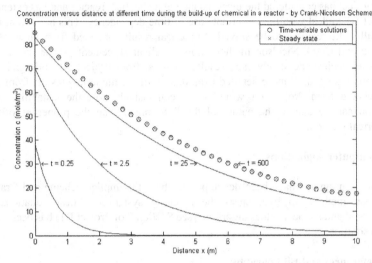

Figure 5.3b. Concentration versus distance at different times during the build-up of chemical in a reactor by Crank-Nicolson scheme.

These results are further compared with the steady state results in Table B5.1 and clearly there is excellent agreement.

Table B5.1. Comparison of concentration results at $t = 500$ using finite difference methods with intervals $\Delta x = 0.25$, $\Delta t = 0.025$ (for $D = 1$, $U = 1$, $\gamma = 0.2$, $c_{in} = 100$, $L = 10$).

x	Implicit Method	Crank-Nicolson method	Steady state solution
0	85.40949391791340	85.40949391791396	85.41021790798811
2.5	55.72990523497404	55.72990523497476	55.72454779633919
5	36.36967495688857	36.36967495688973	36.36263229460928
7.5	23.84430920620443	23.84430920620534	23.84077640806307
10	17.69873908593070	17.69873908593142	17.73340643352621

It should be noted that in such dynamic calculations, the time step is constrained by the stability criterion

$$\Delta t \le \frac{(\Delta x)^2}{2D + \gamma(\Delta x)^2}.$$ (B5.29)

Thus, the reaction terms act to make the time step smaller.

6. Interpretation and Conclusions

This project has considered the mass balance of a chemical reactor and has extended Project B1 to deal with the important case of time dependency. This has meant that the full PDE requires to be solved. Numerical results obtained from an implicit method and Crank-Nicolson method show excellent agreement. These results are validated against the steady state results obtained from Project B1 and again the agreement is good. A more detailed examination of the numerical results from this project and from Project B1 would show conclusively how the instability of a solution can be due to the nature of the PDE rather than the properties of the numerical method.

7. Computer Algorithms

Computer programs have been developed for both the implicit scheme and Crank-Nicolson method. In both cases the resulting system of linear equations is tridiagonal and so the Thomas algorithm (see Section 7 of Project B1) has been used in the solution.

8. References and Bibliography

1. Ames, W.F., *Numerical Methods for Partial Differential Equations*, Academic Press, New York, 1977.
2. Carnahan, B., Luther, H.A. and Wilkes, J.O., *Applied Numerical Methods*, Wiley, New York, 1969.
3. Cussler, E.L., *Diffusion: Mass Transfer in Fluid Systems*, 2nd ed., Cambridge University Press, New York, 1997.
4. Ferziger, J.H., *Numerical Methods for Engineering Application*, Wiley, New York, 1981.
5. Gerald, C.F. and Wheatley, P.O., *Applied Numerical Analysis*, 4th ed., Addison-Wesley, Reading, MA, 1989.
6. Jewell, T.K., *Computer Applications for Engineers*, Wiley, New York, 1991.

Project B6

FLOW THROUGH POROUS MEDIA

SUMMARY: A rectangular plate with fixed boundary conditions is an ideal context for demonstrating how elliptic PDEs can be solved numerically. However, more realistic problems involve geometries with irregular shape. This project considers the flow of liquid through porous media. The geometry to be considered involves a rectangular region with an irregular edge. Values of the head and its partial derivatives are specified on the boundary of the region. A mathematical model is formulated after making some simplifying assumptions. Numerical methods are then used to solve the governing Laplace's equation and including irregular boundary conditions. Consequently numerical results are obtained for the distribution of the head by using MATLAB.

1. Background

Elliptic partial differential equations arise usually from equilibrium or steady-state problems and their solutions frequently maximize or minimize an integral representing the energy of the system. The best known elliptic equations are Poisson's equation

$$\frac{\partial^2 u}{\partial x^2} + \frac{\partial^2 u}{\partial y^2} = f(x, y) \qquad (B6.1)$$

and Laplace's equation

$$\frac{\partial^2 u}{\partial x^2} + \frac{\partial^2 u}{\partial y^2} = 0 . \qquad (B6.2)$$

Poisson's equation represents, for example, the slow motion of incompressible viscous fluid and the inverse square law theories of electricity, magnetism and gravitating matter at points where the charge density, pole strength or mass density, respectively, are non-zero. Laplace's equation arises in the theories associated with the steady flow of heat or electricity in homogeneous conductors, with the irrotational flow of incompressible fluid and with potential problems in electricity, magnetism and gravitating matter at points devoid of these entities.

The domain of integration of a two-dimensional elliptic equation is always an area S bounded by a closed curve C. The boundary condition usually specifies either the value of the function or the value of its normal derivative at every point on C, or a mixture of both. For Laplace's equation it can be shown that the solution throughout S is bounded by the extreme values of u on C and has no maxima or minima.

In this project we will consider the case of fluid flow through porous media which can be described by Laplace's equation. In the numerical solution we will expend our capabilities to address realistic problems involving boundaries at which the derivative is specified and boundaries that are irregularly shaped.

2. Problem Statement

Consider the flow of fluid through porous media. The geometry to be considered is shown in Figure B6.1 and involves a rectangular region with an irregular edge at the bottom right hand corner.

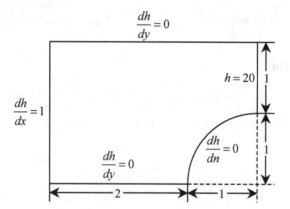

Figure B6.1. Rectangular porous media with an irregular shaped boundary.

Values of the head h and its partial derivatives are as specified on the boundary of the region.

We will use numerical methods to determine the distribution of head for the system shown in Figure B6.1. Any assumptions made in the mathematical model will be clearly stated.

3. Model Formulation

In this project the flow through porous media can be described by Laplace's equation

$$\frac{\partial^2 h}{\partial x^2} + \frac{\partial^2 h}{\partial y^2} = 0, \tag{B6.3}$$

where h is the head. In formulating the mathematical model we make the following assumptions:

A1. The porous media has the same depth everywhere.
A2. The head is fixed at the right hand edge (i.e., $h = 20$).
A3. The irregular shaped boundary at the right bottom corner is a perfect circular arc.
A4. There is no physical factor (e.g., wind) acting which will affect the water flow.

In order to simplify the solution, we use a square mesh of size $\Delta x = \Delta y = 0.5$. Then the important mesh points involved in the determination of the distribution of head for the porous media can be classified into four types:

(1)   given boundary points, i.e., $h = 20$ (indicated by □ in Figure B6.2)
(2)   interior points (indicated by ● in Figure B6.2)
(3)   regular boundary points (indicated by x in Figure B6.2)
(4)   irregular boundary points (indicated by ○ in Figure B6.2).

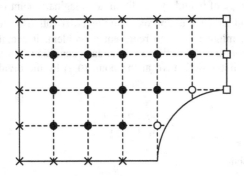

Figure B6.2. Classification of the node points for the porous media.

For points of type (1) h is given. We use for points of type (2) central finite difference approximations

$$\frac{\partial^2 h}{\partial x^2} \cong \frac{h_{i+1,j} - 2h_{i,j} + h_{i-1,j}}{(\Delta x)^2}$$

(B6.4)

and

$$\frac{\partial^2 h}{\partial y^2} \cong \frac{h_{i,j+1} - 2h_{i,j} + h_{i,j-1}}{(\Delta y)^2},$$

(B6.5)

which have errors $O((\Delta x)^2)$ and $O((\Delta y)^2)$, respectively. For points of type (3) and (4) we require to use other techniques.

The case where derivative boundary conditions are given is commonly referred to as Neumann boundary conditions. Figure B6.3 illustrates a node $(0, j)$ at the left edge of the porous media.

Figure B6.3. An imaginary point (-1,*j*) lying outside the boundary with node (0,*j*) at the left edge of the porous media.

To deal with this type of boundary condition, an imaginary point $(-1, j)$ is located a distance Δx beyond the edge to approximate the x-derivative at the edge. Although this exterior point might appear to represent a problem, it actually serves as the vehicle for incorporating the derivative boundary condition into the problem. We can represent the first x-derivative at the point $(0, j)$ by the divided difference to give

$$\frac{\partial h}{\partial x} \cong \frac{h_{1,j} - h_{-1,j}}{2\Delta x}$$

(B6.6)

from which we obtain

$$h_{-1,j} = h_{1,j} - 2\Delta x \frac{\partial h}{\partial x}.$$

(B6.7)

The points of type (4) are more difficult to deal with because they involve derivative conditions for irregularly shaped boundaries. Figure B6.4 shows two such points near the irregular boundary where the normal derivative is specified. The normal

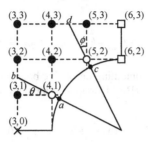

Figure B6.4. A curved boundary where the normal gradient is specified.

derivative at nodes a and c can be approximated by the gradient between nodes (4,1) and b, nodes (5,2) and d, respectively. This gives

$$\left.\frac{\partial h}{\partial n}\right|_a \cong \frac{h_{4,1} - h_b}{L_a} \tag{B6.8}$$

and

$$\left.\frac{\partial h}{\partial n}\right|_c \cong \frac{h_{5,2} - h_d}{L_c}, \tag{B6.9}$$

where L_a is the distance between node (4,1) and b, and L_c is the distance between node (5,2) and d.

4. Numerical Solution

We now use the equations in Section 3 to help us to find a numerical solution for the problem specified in Section 2. First of all, we require some mathematical manipulation of the equations to enable us to formulate a system of linear equations using finite difference techniques.

Using equations (B6.4) and (B6.5), Laplace's equation (B6.3) can be approximated at the general internal mesh point (i, j) by

$$\frac{h_{i+1,j} - 2h_{i,j} + h_{i-1,j}}{(\Delta x)^2} + \frac{h_{i,j+1} - 2h_{i,j} + h_{i,j-1}}{(\Delta y)^2} = 0, \tag{B6.10}$$

where $h_{i,j} = h(i\Delta x, j\Delta y)$, $i = 0,1,2,\cdots,M$, $j = 0,1,2,\cdots,N$.

In our case we will use a square mesh ($\Delta x = \Delta y$) which means that equation (B6.10) simplifies to

$$h_{i+1,j} + h_{i-1,j} + h_{i,j+1} + h_{i,j-1} - 4h_{i,j} = 0. \tag{B6.11}$$

For points on the boundary containing derivative boundary conditions we substitute equation (B6.7) into equation (B6.11) to give at the point $(0, j)$

$$2h_{1,j} - 2\Delta x \frac{\partial h}{\partial x} + h_{0,j+1} - 4h_{0,j} = 0. \tag{B6.12}$$

Similar relationships can be developed for incorporating the derivative boundary conditions at the other edges.

We now return to the points of type (4) mentioned in Section 3. Figure B6.4 shows that when the angles θ and ϕ are less than $45°$, the distance from node (3,1) to b is $\Delta x \tan \theta$ and the distance from node (5,3) to d is $\Delta y \tan \phi$. Linear interpolation can then be used to estimate h_b and h_d in equations (B6.8) and (B6.9). This gives

$$h_b = h_{3,1} + (h_{3,2} - h_{3,1}) \frac{\Delta x \tan \theta}{\Delta y} \tag{B6.13}$$

and

$$h_d = h_{5,3} + (h_{4,3} - h_{5,3}) \frac{\Delta y \tan \phi}{\Delta x}. \tag{B6.14}$$

The lengths L_a and L_b in equations (B6.8) and (B6.9) are equal to $\frac{\Delta x}{\cos \theta}$ and $\frac{\Delta y}{\cos \phi}$, respectively. These lengths, together with the approximations h_b and h_d in equations (B6.13) and (B6.14), respectively, can be substituted back into equations (B6.8) and (B6.9) to give

$$h_{4,1} = \left(\frac{\Delta x}{\cos \theta} \right) \frac{\partial h}{\partial n}\bigg|_a + h_{3,2} \frac{\Delta x \tan \theta}{\Delta y} + h_{3,1} \left(1 - \frac{\Delta x \tan \theta}{\Delta y} \right) \tag{B6.15}$$

and

$$h_{5,2} = \left(\frac{\Delta y}{\cos\phi}\right)\frac{\partial h}{\partial n}\bigg|_c + h_{4,3}\frac{\Delta y\tan\phi}{\Delta x} + h_{5,3}\left(1-\frac{\Delta y\tan\phi}{\Delta x}\right). \tag{B6.16}$$

Such equations provide a means for incorporating the normal derivative into the finite difference approach. Thus, now all the four types of points can be included to determine the distribution of head for the porous media.

The numerical solution then proceeds as follows:

(1) For $i = 0$, $j = 0$, applying equation (B6.11) gives

$$h_{1,0} + h_{-1,0} + h_{0,1} + h_{0,-1} - 4h_{0,0} = 0. \tag{B6.17}$$

Since we have both x- and y- derivative boundary conditions for the left hand corner node point $(0,0)$, then by finite differences we have

$$\frac{\partial h}{\partial x} \cong \frac{h_{1,0} - h_{-1,0}}{2\Delta x} \Rightarrow h_{-1,0} = h_{1,0} - \frac{\partial h}{\partial x} = h_{1,0} - 1 \tag{B6.18}$$

and

$$\frac{\partial h}{\partial y} \cong \frac{h_{0,1} - h_{0,-1}}{2\Delta y} \Rightarrow h_{0,-1} = h_{1,0} - \frac{\partial h}{\partial y} = h_{0,1}. \tag{B6.19}$$

This can be substituted into equation (B6.17) to give

$$2h_{1,0} + 2h_{0,1} - 4h_{0,0} = 1. \tag{B6.20}$$

(2) For $i = 0$, $j = 1$, applying equation (B6.12) gives

$$2h_{1,1} + h_{0,2} + h_{0,0} - 4h_{0,1} = 1. \tag{B6.21}$$

The equations for all the other regular boundary points in Figure 6.2 can be formulated by the same method.

(3) For the two irregular boundary points $(4,1)$ and $(5,2)$, we apply equations (B6.15) and (B6.16) to give

$$h_{4,1} = 0 + \frac{1}{2}h_{3,2} + h_{3,1}\left(1-\frac{1}{2}\right)$$

and

$$h_{5,2} = 0 + \frac{1}{2}h_{4,3} + h_{5,3}\left(1 - \frac{1}{2}\right)$$

since $\tan\theta = \tan\phi = \frac{1}{2}$. This implies that

$$h_{3,1} - 2h_{4,1} + h_{3,2} = 0 \tag{B6.22}$$

and

$$-2h_{5,2} + h_{4,3} + h_{5,3} = 0. \tag{B6.23}$$

Thus, we can now formulate the system of linear equations as

$$
\left[\begin{smallmatrix}
-4 & 2 & & & 2 & & & & & & & & & & & & & & & & & & & & & & \\
1 & -4 & 1 & & & 2 & & & & & & & & & & & & & & & & & & & & & \\
& 1 & -4 & 1 & & & 2 & & & & & & & & & & & & & & & & & & & & \\
& & & -1 & & & & 1 & & & & & & & & & & & & & & & & & & & \\
1 & & & & -4 & 2 & & & & 1 & & & & & & & & & & & & & & & & & \\
& 1 & & & 1 & -4 & 1 & & & & 1 & & & & & & & & & & & & & & & & \\
& & 1 & & & 1 & -4 & 1 & & & & 1 & & & & & & & & & & & & & & & \\
& & & 1 & & & 1 & -4 & 1 & & & & 1 & & & & & & & & & & & & & & \\
& & & & & & & -1 & -1 & & & & 2 & & & & & & & & & & & & & & \\
& & & & 1 & & & & & -4 & 2 & & & & & 1 & & & & & & & & & & & \\
& & & & & 1 & & & & 1 & -4 & 1 & & & & & 1 & & & & & & & & & & \\
& & & & & & 1 & & & & 1 & -4 & 1 & & & & & 1 & & & & & & & & & \\
& & & & & & & 1 & & & & 1 & -4 & 1 & & & & & 1 & & & & & & & & \\
& & & & & & & & 1 & & & & 1 & -4 & 1 & & & & & 1 & & & & & & & \\
& & & & & & & & & & & & & -1 & & & & & & 2 & -1 & & & & & & \\
& & & & & & & & & 1 & & & & & & -4 & 2 & & & & & 1 & & & & & \\
& & & & & & & & & & 1 & & & & & 1 & -4 & 1 & & & & & 1 & & & & \\
& & & & & & & & & & & 1 & & & & & 1 & -4 & 1 & & & & & 1 & & & \\
& & & & & & & & & & & & 1 & & & & & 1 & -4 & 1 & & & & & 1 & & \\
& & & & & & & & & & & & & 1 & & & & & 1 & -4 & 1 & & & & & 1 & \\
& & & & & & & & & & & & & & 1 & & & & & 1 & -4 & & & & & & 1 \\
& & & & & & & & & & & & & & & 2 & & & & & & -4 & 2 & & & & \\
& & & & & & & & & & & & & & & & 2 & & & & & 1 & -4 & 1 & & & \\
& & & & & & & & & & & & & & & & & 2 & & & & & 1 & -4 & 1 & & \\
& & & & & & & & & & & & & & & & & & 2 & & & & & 1 & -4 & 1 & \\
& & & & & & & & & & & & & & & & & & & 2 & & & & & 1 & -4 & 1 \\
& & & & & & & & & & & & & & & & & & & & 2 & & & & & 1 & -4
\end{smallmatrix}\right]
\begin{bmatrix}
h_{0,0} \\ h_{1,0} \\ h_{2,0} \\ h_{3,0} \\ h_{0,1} \\ h_{1,1} \\ h_{2,1} \\ h_{3,1} \\ h_{4,1} \\ h_{0,2} \\ h_{1,2} \\ h_{2,2} \\ h_{3,2} \\ h_{4,2} \\ h_{5,2} \\ h_{0,3} \\ h_{1,3} \\ h_{2,3} \\ h_{3,3} \\ h_{4,3} \\ h_{5,3} \\ h_{0,4} \\ h_{1,4} \\ h_{2,4} \\ h_{3,4} \\ h_{4,4} \\ h_{5,4}
\end{bmatrix}
=
\begin{bmatrix}
1 \\ 0 \\ 0 \\ 0 \\ 1 \\ 0 \\ 0 \\ 0 \\ 0 \\ 1 \\ 0 \\ 0 \\ 0 \\ 0 \\ 0 \\ 1 \\ 0 \\ 0 \\ 0 \\ 0 \\ -20 \\ 1 \\ 0 \\ 0 \\ 0 \\ 0 \\ -20
\end{bmatrix}
$$

$$\tag{B6.24}$$

By solving the above linear system of equations, we obtain the values of the head $h_{i,j}$ at the 27 node points (i, j) where $i = 0,1,\cdots,4$, $j = 0,1,\cdots,4$ [with exception of $(4,0)$] and the points $(5,2)$, $(5,3)$, $(5,4)$. The calculated values of the water flow velocity at all the above node points are presented in Figure B6.5.

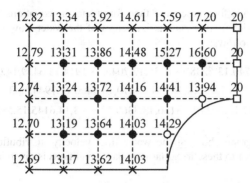

Figure B6.5. Water flow velocity distribution through porous media subject to irregular boundary conditions.

4.1 MATLAB APPROACH

For the porous media with irregular boundary, we should first determine which node points need to be considered and which can be ignored. This can be done by calculating the distance from the point to the central point of the circular arc. Since the irregular boundary is a circular arc with radius 1 unit, then if this distance is less than or equal to 1 this point need not be considered.

The algorithm for this MATLAB approach is given in Section 7. When this procedure is applied with the testing data ($\Delta x = \Delta y = 0.5$), the index (x, y) will store the values as follows:

$$
\begin{array}{|c|c|c|c|c|}
\hline
1 & 5 & 10 & 16 & 22 \\
\hline
2 & 6 & 11 & 17 & 23 \\
\hline
3 & 7 & 12 & 18 & 24 \\
\hline
4 & 8 & 13 & 19 & 25 \\
\hline
0 & 9 & 14 & 20 & 26 \\
\hline
0 & 0 & 15 & 21 & 27 \\
\hline
0 & 0 & 15 & 21 & 27 \\
\hline
\end{array}
$$

From the above results, 0 means that the point is outside the porous media (i.e., within the circular arc of radius 1) and so can be ignored. We also see that 15, 21 and 27 are duplicated because these three points lie on the top right hand boundary, which means that they need not be considered. We simply set the values to the given boundary condition ($h = 20$).

Once the computer knows which points should be considered, it can classify them into one of the above four types and then formulate the equations for each point. After finding the global stiffness matrix, values of the water flow velocity **v** have been calculated as follows:

$$\mathbf{v} = (12.6893, 13.1740, 13.6238, 14.0336, 12.7046, 13.1914, 13.6439, 14.0336, 14.2934,$$

$$12.7463, 13.2131, 13.7268, 14.1635, 14.4170, 13.9400, 12.7944, 13.3078, 13.8568,$$

$$14.4765, 15.2712, 16.6024, 12.8156, 13.3369, 13.9163, 14.6145, 15.5888, 17.198)^T$$

A 3-dimensional graph showing the water flow velocity distribution through the porous media subject to these irregular boundary conditions is given in Figure B6.6.

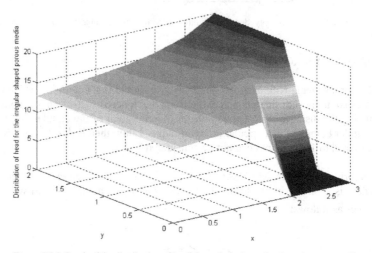

Figure B6.6. Graph of the distribution of head through the irregular shaped porous media.

5. Model Validation

It is important to validate the numerical results obtained in Section 4. This has been done by using another computer program.

For the derivative boundary condition $\dfrac{\partial h}{\partial x} = 1$, applying equation (B6.12) gives

$$h_{0,j} = \frac{1}{4}\left(2h_{1,j} - 2\Delta x \frac{\partial h}{\partial x} + h_{0,j+1} + h_{0,j-1}\right). \tag{B6.25}$$

Similar relationships can be developed for derivative boundary conditions $\dfrac{\partial h}{\partial y} = 0$ at the other edges to give

$$h_{i,0} = \frac{1}{4}\left(2h_{i,1} - 2\Delta y\frac{\partial h}{\partial y} + h_{i+1,0} + h_{i-1,0}\right) \qquad \text{(B6.26)}$$

and

$$h_{i,4} = \frac{1}{4}\left(2h_{i,3} + 2\Delta y\frac{\partial h}{\partial y} + h_{i+1,4} + h_{i-1,4}\right). \qquad \text{(B6.27)}$$

For the irregular boundary condition $\frac{\partial h}{\partial n} = 0$, applying equations (B6.22) and (B6.23) gives

$$h_{4,1} = \frac{1}{2}h_{3,2} + \frac{1}{2}h_{3,1} \qquad \text{(B6.28)}$$

and

$$h_{5,2} = \frac{1}{2}h_{4,3} + \frac{1}{2}h_{5,3}. \qquad \text{(B6.29)}$$

Now we have the equations to deal with all the interior points, the derivative boundary points and the irregular boundary points. By running the program using the above equations, we calculate the values of the elements of the head velocity matrix H, namely,

$$H = \begin{bmatrix} h_{0,4} & h_{1,4} & h_{2,4} & h_{3,4} & h_{4,4} & h_{5,4} & h_{6,4} \\ h_{0,3} & h_{1,3} & h_{2,3} & h_{3,3} & h_{4,3} & h_{5,3} & h_{6,3} \\ h_{0,2} & h_{1,2} & h_{2,2} & h_{3,2} & h_{4,2} & h_{5,2} & h_{6,2} \\ h_{0,1} & h_{1,1} & h_{2,1} & h_{3,1} & h_{4,1} & h_{5,1} & h_{6,1} \\ h_{0,0} & h_{1,0} & h_{2,0} & h_{3,0} & h_{4,0} & h_{5,0} & h_{6,0} \end{bmatrix} \qquad \text{(B6.30)}$$

After the first iteration, the results are

$$H = \begin{bmatrix} -0.4160 & -0.1587 & -0.0575 & -0.0201 & -0.0072 & 7.4966 & 20 \\ -0.3320 & -0.1094 & -0.0356 & -0.0114 & -0.0043 & 4.9968 & 20 \\ -0.3281 & -0.1055 & -0.0332 & -0.0100 & -0.0058 & -0.0086 & 20 \\ -0.3125 & -0.0938 & -0.0273 & -0.0068 & -0.0132 & 0 & 0 \\ -0.2500 & -0.0625 & -0.0156 & 0.0000 & 0 & 0 & 0 \end{bmatrix}$$

After two iterations, we have

$$H = \begin{bmatrix} -0.5972 & -0.2813 & -0.1237 & -0.0578 & 2.4776 & 8.8942 & 20 \\ -0.5356 & -0.2353 & -0.0966 & -0.0381 & 1.2328 & 6.5496 & 20 \\ -0.5078 & -0.2144 & -0.0823 & -0.0313 & -0.0203 & -2.5312 & 20 \\ -0.4883 & -0.1953 & -0.0723 & -0.0256 & -0.0370 & 0 & 0 \\ -0.4375 & -0.1602 & -0.0537 & -0.0068 & 0 & 0 & 0 \end{bmatrix}$$

The sequence of iterations is continued until the maximum percentage relative difference between successive values is 0.001%. On this basis we find that after 693 iterations, we have

$$H = \begin{bmatrix} 12.8040 & 13.3255 & 13.9055 & 14.6047 & 15.5807 & 17.1932 & 20 \\ 12.7827 & 13.2963 & 13.8459 & 14.4664 & 15.2624 & 15.5961 & 20 \\ 12.7344 & 13.2314 & 13.7156 & 14.1528 & 14.4067 & 13.9288 & 20 \\ 12.6926 & 13.1796 & 13.6324 & 14.0226 & 14.2829 & 0 & 0 \\ 12.6772 & 13.1621 & 13.6123 & 14.0225 & 0 & 0 & 0 \end{bmatrix}$$

On comparing these results with MATLAB results in Section 4, the maximum percentage difference is 0.0954% which is small and hence give us confidence in the accuracy of the results.

6. Interpretation and Conclusions

This project has dealt with the realistic problem of flow of liquid through porous media with irregular shaped boundary. Boundary conditions involving derivatives have been dealt with on both regular and irregular sections of the boundary. Good agreement has been obtained from numerical results obtained by MATLAB and other programs.

This project could be extended by realizing that the velocity of water flow through the porous media can be related to head by D'Arcy's law

$$q_n = -\kappa \frac{dh}{dn},$$

where κ is the hydraulic conductivity and q_n is the discharge velocity in the normal n direction. So, in this way, it would be possible to compute the water velocity for the problem for particular values of κ.

7. Computer Algorithms

The computer algorithm for the MATLAB approach in Section 4.1 is given below:

```
Initialize a index(x,y) array
Set count = 0
for j = 1 to y
      for i = 1 to x
            Calculate distance of point(i,j) to the right bottom corner
            if distance > 1 (i.e. outside the irregular boundary)
                  if not right boundary
                        Increase count by 1
                  end
                  Set index(i,j) = count
            end
            if distance ≥ 1 and point at right boundary
                  Set index(i,j) = count
            end
      end
end
```

8. References and Bibliography

1. Albertson, M.L., Barton, J.R. and Simons, D.B., *Fluid Mechanics for Engineers*, Prentice Hall, Englewood Cliffs, N.J., 1960.
2. Carnahan, B., Luther, H.A. and Wilkes, J.O., *Applied Numerical Methods*, Wiley, New York, 1969.
3. Copson, E.T., *Partial Differential Equations*, Cambridge University Press, London, 1975.
4. Huebner, K.H., Thornton, E.A. and Byrom, T.G., *The Finite Element Method for Engineers*, Wiley, New York, 1995.
5. Jaluria, Y., *Computer Methods for Engineering*, Allyn and Bacon, Inc., Boston, 1988.
6. Johnson, C., *Numerical Solution of Partial Differential Equations by the Finite Element Method*, Cambridge University Press, Cambridge, 1987.
7. Köckler, N., *Numerical Methods and Scientific Computing*, Oxford University Press, New York, 1994.
8. Rice, J.R., *Numerical Methods, Software and Analysis*, Academic Press, London, 1983.

PDE Problems

1. Write a computer program to determine the numerical solution of Laplace's equation

$$\frac{\partial^2 u}{\partial x^2} + \frac{\partial^2 u}{\partial y^2} = 0$$

and Poisson's equation

$$\frac{\partial^2 u}{\partial x^2} + \frac{\partial^2 u}{\partial y^2} = f,$$

for a rectangular object of variable width and height. The object could have Dirichlet, Neumann or Cauchy boundary conditions. The value of f in Poisson's equation should be assumed constant. Use this program to find the solution of the following problems:

(a) A thin metal plate of dimension 2ft × 2ft is subjected to four heat sources which maintain the temperatures on its four edges as follows:

$$u(x,0) = 400 \,^\circ C, \; u(0,y) = 200 \,^\circ C,$$

$$u(x,2) = 50 \,^\circ C, \; u(2,y) = 100 \,^\circ C.$$

The flat sides of the plate are insulated so that no heat is transferred through these sides. Calculate the temperature profiles within the plate.

(b) Perfect insulation is installed on two edges (right and top) of the plate of part (a). The other two edges are exposed to heat sources. This means that the set of Dirichlet and Neumann boundary conditions is

$$u(x,0) = 400 \ {}^\circ C, \ u(0,y) = 200 \ {}^\circ C,$$

$$\left.\frac{\partial u}{\partial y}\right|_{x,2} = 0, \ \left.\frac{\partial u}{\partial x}\right|_{2,y} = 0.$$

Calculate the temperature profiles within the plate and compare these with the results from part (a).

(c) The thin metal plate of part (a) is made of an alloy which has a melting point of $600\ {}^\circ C$ and a thermal conductivity of 15 Btu/(hour. ft. ${}^\circ C$). The plate is subjected to an electric current which creates a uniform heat source within the plate. The amount of heat generated is $Q = 100,000$ Btu/(hour. ft^3). The edges of the plate are in contact with heat sinks which maintain the temperature $50\ {}^\circ C$ on all four edges. Examine the temperature profiles within the plate to ascertain whether the alloy will begin to melt under these conditions.

(d) Determine the optimum value of the overrelaxation parameter for the conditions used in part (a).

2. Modify the computer program in Problem 1 to solve the three-dimensional problem

$$\frac{\partial^2 u}{\partial x^2} + \frac{\partial^2 u}{\partial y^2} + \frac{\partial^2 u}{\partial z^2} = 0.$$

Apply this program to calculate the distribution of the dependent variable within a solid body which is subjected to the following boundary conditions:

$$u(0,y,z) = 200, \ u(2,y,z) = 200,$$

$$u(x,0,z) = 0, \ u(x,2,z) = 0,$$

$$u(x,y,0) = 100, \ u(x,y,2) = 100.$$

3. (a) Solve Laplace's equation with the following boundary conditions

$$u(0, y) = 80, \quad \frac{\partial u}{\partial x}\bigg|_{5,y} = 20,$$

$$\frac{\partial u}{\partial y}\bigg|_{x,0} = 0, \quad \frac{\partial u}{\partial y}\bigg|_{x,1} = 0.$$

Discuss the results and determine the optimum value of the overrelaxation parameter for this problem.

(b) Extend the computer program in (a) to include Robbins boundary conditions of the form:

$$k\frac{\partial u}{\partial x} = h(u - u_f) \text{ at } x = 0 \text{ and } t \geq 0,$$

$$-k\frac{\partial u}{\partial x} = h(u - u_f) \text{ at } x = C \text{ and } t \geq 0,$$

where u is the value of the dependent variable at the boundary and u_f is a known value of the dependent variable in the fluid next to the boundary; k, h and C are known constants.

Apply this program to solve the following problem: The ambient temperature surrounding a house is $60°F$. The heating in the house has been turned off and so the internal temperature is also $60°F$ at t = 0. The heating system is turned on and raises the internal temperature to $75°F$ at the rate of $5°F$ /hour. The ambient temperature remains at $60°F$. The wall of the house is 0.4 ft thick and is made of material which has an average thermal diffusivity $\alpha = 0.01 ft^2$/hour and a thermal conductivity $k = 0.2$ Btu/(hour. ft. $°F$). The heat transfer coefficient on the inside of the wall is $h_{in} = 1.2$Btu/(hour. ft². $°F$) and the heat transfer coefficient on the outside is $h_{out} = 2.0$Btu/(hour. ft². $°F$). Estimate how long it will take to reach a steady-state temperature distribution across the wall.

4. Develop the finite difference approximation of Fick's second law of diffusion in polar coordinates, namely

$$\frac{\partial c}{\partial t} = D\left(\frac{\partial^2 c}{\partial r^2} + \frac{1}{r}\frac{\partial c}{\partial r} + \frac{1}{r^2}\frac{\partial^2 c}{\partial \theta^2} + \frac{\partial^2 c}{\partial z^2}\right),$$

where $c(r,\theta,z)$ represents the concentration and D the diffusivity. Hence write a computer program which can be used to solve the following problem:

A wet cylinder of agar gel at 278 $^\circ K$ with a uniform concentration of urea of 0.1 kg. mol/m^3 has a diameter of 3cm and is 4cm long with flat parallel ends. The diffusivity is 4.5×10^{-10} m^2/s. Calculate the concentration at the midpoint of the cylinder after 100 hours for the following cases if the cylinder is suddenly immersed in turbulent pure water:

(a) Radial diffusion only.
(b) Diffusion that occurs radially and axially.

5. Consider a first-order chemical reaction being carried out under isothermal steady-state conditions, in a tubular-flow reactor. On the assumptions of laminar flow and negligible axial diffusion, the material balance equation is

$$-v_0\left[1-\left(\frac{r}{R}\right)^2\right]\frac{\partial c}{\partial z} + D\left(\frac{\partial^2 c}{\partial r^2} + \frac{1}{r}\frac{\partial c}{\partial r}\right) - kc = 0,$$

where v_0 = velocity of central stream line
R = tube radius
k = reaction-velocity constant
D = radial diffusion constant
c = concentration of reactant
z = axial distance down the tube
r = radial distance from the centre.

After defining the following dimensionless variables:

$$\alpha = \frac{kz}{v_0}, \ C = \frac{c}{c_0}, \ \beta = \frac{D}{kR^2}, \ V = \frac{r}{R},$$

the equation becomes

$$(1-V^2)\frac{\partial C}{\partial \alpha} = \beta\left(\frac{\partial^2 C}{\partial V^2} + \frac{1}{V}\frac{\partial C}{\partial V}\right) - C,$$

where c_0 is the entering concentration of the reactant to the reactor.

(a) Choose a set of appropriate boundary conditions for this problem and explain your choice.

(b) What class of PDE is the above equation (hyperbolic, parabolic or elliptic)?

(c) Set up the equation for numerical solution using finite difference approximations.

(d) Does your choice of finite differences result in an explicit or implicit set of equations? Give details of the procedure for the solution of this set of equations.

(e) Discuss stability considerations with respect to the method you have chosen.

6. A square membrane of side 12in (no bending or shear stresses), with a square hole of side 3in in the middle, is fastened at the inside and outside boundaries as shown in Figure 1. If a highly stretched membrane is subjected to a pressure p, the PDE for the deflection u in the z-direction is

$$\frac{\partial^2 u}{\partial x^2} + \frac{\partial^2 u}{\partial y^2} = -\frac{p}{T},$$

where T is the tension (lb/in). For a highly stretched membrane, the tension T may be assumed constant for small deflections. Assume the following values of pressure and tension:

$$p = 10\text{lb/in}^2 \text{ (uniformly distributed)}$$

$$T = 200\text{lb/in} .$$

(a) Express the differential equation in finite difference form to obtain the deflections u of the membrane.

(b) List all the boundary conditions needed for the numerical solution of the problem. Utilize some or all of these boundary conditions to simplify the finite difference equations in part (a).

(c) Develop a computer program for the solution of this problem.

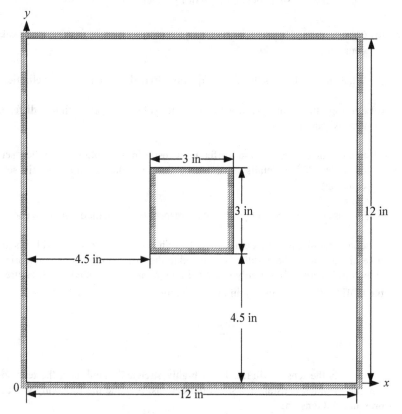

Figure 1. Stretched membrane fastened at the inside and outside boundaries.

7. Figure 2 shows the cross section of a long cooling fin of thickness t, width w and thermal conductivity k which is bonded to a hot wall, maintaining its base (at $x = 0$) at a temperature T_0. Heat is conducted steadily through the fin in the plane of Figure 2 so that the fin temperature T obeys Laplace's equation

$$\frac{\partial^2 T}{\partial x} + \frac{\partial^2 T}{\partial y^2} = 0.$$

Temperature variations along the length of the fin in the z-direction are ignored.

Heat is lost from the sides and tip of the fin by convection to the surrounding air (radiation is neglected at sufficiently low temperatures) at a local rate

$$Q = h(T_s - T_a) \text{ Btu/(hour.ft}^2).$$

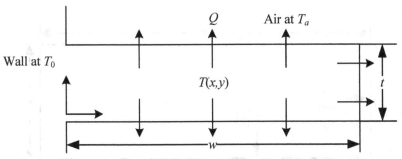

Figure 2. Cross section of a long cooling fin.

Here, T_s and T_a (in degrees Fahrenheit) are the temperatures at a point on the fin surface and of the air, respectively. If the surface of the fin is vertical, the heat transfer coefficient h obeys the dimensional correlation

$$h = 0.2(T_s - T_a)^{\frac{1}{3}}.$$

(a) Set up the equation for a numerical solution of this problem to determine the temperature at a finite number of points within the fin and at the surface of the fin.

(b) Describe in detail the step-by-step procedure for solving the equation in part (a) and evaluating the temperatures within the fin and at the surface.

(c) Develop a computer program to find the numerical solution of this problem using the following data:

$$t = 0.5\text{in}, \; k = 26 \text{ Btu/(hour. ft. } ^\circ F),$$
$$T_w = 210^\circ F, \; T_a = 60^\circ F$$
$$w = 1.0\text{in}.$$

8. Compute the steady-state distribution of concentration for the tank shown in Figure 3. The PDE governing this system is

$$D\left(\frac{\partial^2 c}{\partial x^2} + \frac{\partial^2 c}{\partial y^2} \right) - kc = 0$$

and the boundary conditions are as shown. Employ a value of 0.2 for D and 0.1 for k.

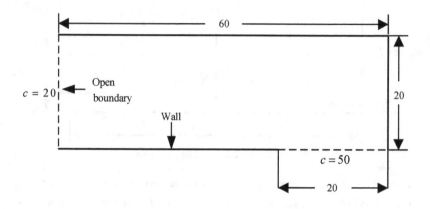

Figure 3. Dimensions and distribution of concentration of a tank.

9. An insulated composite rod is formed of two parts arranged end to end, with both halves of equal length. Part a has thermal conductivity k_a for $0 \leq x \leq 1/2$, and part b has thermal conductivity k_b for $1/2 \leq x \leq 1$.

The nondimensional transient heat conduction equations which describe the temperature T over the length x of the composite rod are

$$\frac{\partial^2 T}{\partial x^2} = \frac{\partial T}{\partial t}, \ 0 \leq x \leq 1/2,$$

$$r\frac{\partial^2 T}{\partial x^2} = \frac{\partial T}{\partial t}, \ 1/2 \leq x \leq 1,$$

where $T =$ temperature, $x =$ axial coordinate, $t =$ time and $r = k_a/k_b$. The boundary conditions are

$$T(0,t) = 1, \ T(1,t) = 1,$$

$$\left(\frac{\partial T}{\partial x}\right)_a = \left(\frac{\partial T}{\partial x}\right)_b, \ x = \frac{1}{2},$$

and the initial condition is

$$T(x,0) = 0, \ 0 < x < 1.$$

Solve this set of equations for the temperature distribution as a function of time. Use second-order finite difference approximations for the derivatives with a Crank-Nicolson formulation to integrate in time. Develop a computer program for the solution and choose values of Δx and Δt to ensure high accuracy. Plot the temperature T versus length x for a range of values of time t. Generate separate curves for the following values of the parameter r:

$$r = 1, 0.1, 0.01, 0.001 \text{ and } 0.$$

10. Solve the nondimensional transient heat conduction equation in two dimensions, which represents the transient temperature distribution in an insulated plate. The governing equation is

$$\frac{\partial^2 T}{\partial x^2} + \frac{\partial^2 T}{\partial y^2} = \frac{\partial T}{\partial t},$$

where $T =$ temperature, x and y are spatial coordinates and $t =$ time. The boundary conditions are

$$T(x,0,t) = 0, \ T(x,1,t) = 1,$$

$$T(0,y,t) = 0, \ T(1,y,t) = 1,$$

and the initial condition is

$$T(x,y,0) = 0, \ 0 \le x < 1, \ 0 \le y < 1.$$

Solve using the alternating direction-implicit (ADI) technique. Develop a computer program to implement the solution. Plot the results using a three-dimensional plotting routine where the horizontal plane contains the x and y axes and the z axis is the dependent variable T. Construct several plots at various times, including the following:

(a) the initial conditions;
(b) one intermediate time, approximately halfway to steady state;
(c) the steady-state condition.

Part III

OPTIMIZATION

Case Study A5

LINEAR PROGRAMMING PROBLEM INVOLVING WINE PRODUCTION

SUMMARY: This case study involves the formulation of a wine production problem as a linear programming problem. A vintner producing two types of wine (M and D) to sell to the local shop knows the profit figures ($/gal) for each type. The requirements of each type of wine in terms of the ingredients, namely, grapes, sugar and extract are also known. As the vintner has some constraints on these ingredients, he wishes to know how best to proceed. A mathematical solution is obtained using the simplex method and sensitivity analysis is used to study the effects of changes in the key parameters on the optimal solution. In this way the vintner obtains important information on how to use his resources to maximize profit. The solution is validated by using the linear programming computer package LINDO and the mathematical software package MAPLE.

1. Background

One of the most important tools of optimization is "linear programming" (L.P.). A linear programming problem is specified by a linear, multi-variable function which is to be optimized (maximized or minimized) subject to a number of linear constraints. The mathematician, G. B. Dantzig [3] developed an algorithm called the "simplex method" to solve problems of this type. The original simplex method has been modified into an efficient algorithm to solve large L.P. problems by computer. Problems from a wide variety of fields can be formulated and solved by means of L.P. This includes resource allocation problems in government planning, network analysis for urban and regional planning, production planning problems in industry and the management of transportation distributive systems. Hence L.P. is one of the successes of modern optimization theory. The mathematical structure of L.P. also allows important questions to be answered concerning the sensitivity of the optimum solution to data changes.

This case study involves the production of two types of wine by a local vintner with the purpose of selling it to the local shop. He knows the profit ($/gal) for both types, i.e., medium white (M) and dry white (D). Production of the wine requires a combination of grapes, sugar and extract. The exact requirements are known for both M and D. The constraints for this problem are given by the limitations for grapes, sugar and extract. Obviously, the objective here is that the vintner should maximize his profit in selling to the local shop. Clearly, there is sufficient information to formulate in mathematical form a linear programming problem. This L.P. problem is solved using the simplex method. Also, sensitivity analysis is employed to examine the effects of changes in the parameters on the optimal solution. The solution is validated by using LINDO and MAPLE.

2. Problem Statement

A local wine producer makes two types of wine, medium white (M) and dry white (D), to sell to the local shop. He makes $5 profit per gallon from M and $4 a gallon from D. Now M requires 3 boxes of grapes, 4 lb of sugar and 2 pints of extract per gallon. Also, D requires 4 boxes of grapes, 2 lb of sugar and 1 pint of extract per gallon. The vintner has 14 boxes of grapes, 8 lb of sugar and 6 pints of extract left before selling his business. We wish to decide how to use these resources to maximize profit.

(a) We will create and solve the dual linear programming problem. Then we will find the optimal solution to the primal problem by interpreting the optimal dual tableau.

(b) By performing sensitivity analysis we will determine for what range of profit for dry white wine the present optimal basis remains optimal.

(c) Suppose the wine producer wishes to vary the supply of grapes he requires in the production of his two white wines. He wants to know if his wine-making business will still be profitable if for some reason there is a shortage of grapes. We will then determine how much below 14 boxes the supply can drop for the present basis to be still optimal.

(d) We return to the original problem but suppose now that the medium white wine requires $7\frac{1}{2}$ units of extract. We will use sensitivity analysis to determine how this affects the solution.

3. Model Formulation

Let x_1 (gal) and x_2 (gal) be the amount of medium white wine (M) and dry white wine (D), respectively. In mathematical form, the problem is:

Primal   Maximize: $5x_1 + 4x_2 = x_0$

subject to: $3x_1 + 4x_2 \leq 14$

$4x_1 + 2x_2 \leq 8$

$2x_1 + x_2 \leq 6$

$x_1 \geq 0$

$x_2 \geq 0.$

Since the number of constraints is greater than the number of variables, the problem is more easily solved when its dual is created. The problem can be written as follows:

Dual   Minimize: $14y_1 + 8y_2 + 6y_3 = y_0'$

subject to: $3y_1 + 4y_2 + 2y_3 \geq 5$

$4y_1 + 2y_2 + y_3 \geq 4$

$y_1, y_2, y_3 \geq 0.$

In standard form the new problem is

Maximize: $-14y_1 - 8y_2 - 6y_3 - My_5 - My_7 = y_0$

subject to: $3y_1 + 4y_2 + 2y_3 - y_4 + y_5 = 5$

$4y_1 + 2y_2 + y_3 - y_6 + y_7 = 4$

$y_i \geq 0,\ i = 1, 2, \cdots, 7,$

where we have introduced slack variables y_4, y_5, y_6, y_7.

4. Mathematical/Numerical Solution

(a) The tableaux required to solve the problem are displayed next

	y_1	y_2	y_3	y_4	y_5	y_6	y_7	r.h.s.
	3	4	2	-1	1	0	0	5
	4	2	1	0	0	-1	1	4
y_0	14	8	6	0	M	0	M	0

	y_1	y_2	y_3	y_4	y_5	y_6	y_7	r.h.s.	Ratio
	3	4	2	-1	1	0	0	5	$\frac{5}{3}$
	$\boxed{4}$	2	1	0	0	-1	1	4	$\frac{4}{4}$
y_0	$-(7M-14)$	$-(6M-8)$	$-(3M-6)$	M	0	M	0	$-9M$	

	y_1	y_2	y_3	y_4	y_5	y_6	y_7	r.h.s.	Ratio
	0	$\boxed{\frac{5}{2}}$	$\frac{5}{4}$	-1	1	$\frac{3}{4}$	$-\frac{3}{4}$	2	$\frac{4}{5}$
	1	$\frac{1}{2}$	$\frac{1}{4}$	0	0	$-\frac{1}{4}$	$\frac{1}{4}$	1	2
y_0	0	$-(\frac{5}{2}M-1)$	$-(\frac{5}{2}M-\frac{5}{2})$	M	0	$-(\frac{3}{4}M-\frac{7}{2})$	$\frac{7M-14}{4}$	$-(2M+14)$	

	y_1	y_2	y_3	y_4	y_5	y_6	y_7	r.h.s.
	0	1	$\frac{1}{2}$	$-\frac{2}{5}$	$\frac{2}{5}$	$\frac{3}{10}$	$-\frac{3}{10}$	$\frac{4}{5}$
	1	0	0	$\frac{1}{5}$	$-\frac{1}{5}$	$-\frac{2}{5}$	$\frac{2}{5}$	$\frac{3}{5}$
y_0	0	0	2	$\frac{2}{5}$	$\frac{10M-4}{10}$	$\frac{16}{5}$	$\frac{5M-8}{5}$	$-\frac{74}{5}$

Hence the solution to the original minimization problem is

$$y_1^* = \frac{3}{5}, \; y_2^* = \frac{4}{5},$$
$$y_i^* = 0, \text{ otherwise,}$$
$$y_0^* = \frac{74}{5} = 14\frac{4}{5}.$$

The solution to the primal problem can be found by observing the slack variables y_4 and y_6, in the objective function row. Thus x_1^* has value $\frac{2}{5}$ and x_2^* has value $\frac{16}{5}$. The wine producer should produce $\frac{2}{5}$ gallon of the medium white wine and $3\frac{1}{5}$ gallons of the dry white wine. He would then maximize his profit at $ 14.80.

(b) In this part an objective function coefficient has been changed. Here we examine the effect that this has on the optimal solution and its value by solving the original problem and then performing sensitivity analysis.

Consider the primal L.P. problem of (a):

Maximize: $5x_1 + 4x_2$

subject to: $3x_1 + 4x_2 \leq 14$
$4x_1 + 2x_2 \leq 8$
$2x_1 + x_2 \leq 6$
$x_1, x_2 \geq 0.$

The problem now becomes:

Maximize: $5x_1 + 4x_2$
subject to: $3x_1 \;+\; 4x_2 \;+\; x_3 \qquad\qquad\qquad = 14$
$4x_1 \;+\; 2x_2 \qquad\;\; +\; x_4 \qquad\qquad = 8$
$2x_1 \;+\; x_2 \qquad\qquad\qquad +\; x_5 \;= 6$
$x_i \geq 0,\, i = 1, 2, \cdots, 5.$

This problem is now solved using the simplex method

	x_1	x_2	x_3	x_4	x_5	r.h.s.	Ratio
	3	4	1	0	0	14	$\frac{14}{3}$
	4	2	0	1	0	8	2
	2	1	0	0	1	6	3
x_0	−5	−4	0	0	0	0	

	x_1	x_2	x_3	x_4	x_5	r.h.s.	Ratio
	0	$\frac{5}{2}$	1	$-\frac{3}{4}$	0	8	$\frac{16}{5}$
	1	$\frac{1}{2}$	0	$\frac{1}{4}$	0	2	4
	0	0	0	$-\frac{1}{2}$	1	2	
x_0	0	$-\frac{3}{2}$	0	$\frac{5}{4}$	0	10	

	x_1	x_2	x_3	x_4	x_5	r.h.s.
	0	1	$\frac{2}{5}$	$-\frac{3}{10}$	0	$\frac{16}{5}$
	1	0	$-\frac{1}{5}$	$\frac{2}{5}$	0	$\frac{2}{5}$
	0	0	0	$-\frac{1}{2}$	1	2
x_0	0	0	$\frac{3}{5}$	$\frac{4}{5}$	0	$14\frac{4}{5}$

Suppose c_2 is changed from 4 to $4 + p$. Then the initial simplex tableau for the problem becomes

	x_1	x_2	x_3	x_4	x_5	r.h.s.
	3	4	1	0	0	14
	4	2	0	1	0	8
	2	1	0	0	1	6
x_0	−5	−(4 + p)	0	0	0	0

The corresponding tableau from this table would be

	x_1	x_2	x_3	x_4	x_5	r.h.s.
	0	1	$\frac{2}{5}$	$-\frac{3}{10}$	0	$\frac{16}{5}$
	1	0	$-\frac{1}{5}$	$\frac{2}{5}$	0	$\frac{2}{5}$
	0	0	0	$-\frac{1}{2}$	1	2
x_0	0	$-p$	$\frac{3}{5}$	$\frac{4}{5}$	0	$14\frac{4}{5}$

In order for the present basis to remain optimal, x_2 must still be basic. Therefore the x_2 value in the x_0 row must have zero value. This results in the following tableau:

	x_1	x_2	x_3	x_4	x_5	r.h.s.
	0	1	$\frac{2}{5}$	$-\frac{3}{10}$	0	$\frac{16}{5}$
	1	0	$-\frac{1}{5}$	$\frac{2}{5}$	0	$\frac{2}{5}$
	0	0	0	$-\frac{1}{2}$	1	2
x_0	0	0	$\frac{3}{5}+\frac{2}{5}p$	$\frac{4}{5}-\frac{3}{10}p$	0	$\frac{74}{5}-\frac{16}{5}p$

For the present basis to remain optimal all x_0 row values must be non-negative. Thus

$$\frac{3}{5}+\frac{2}{5}p \geq 0$$

$$\frac{4}{5}-\frac{3}{10}p \geq 0.$$

This implies

$$-\frac{3}{2} \leq p \leq \frac{8}{3}.$$

Hence the range for c_2 is

$$(4-\tfrac{3}{2},\ 4+\tfrac{8}{3}) = (\tfrac{5}{2}, \tfrac{20}{3}).$$

(c) Again we solve the L.P. problem by the simplex method and analyze the effect of changing a r.h.s. constant.

Consider the problem:

Maximize: $4x_1 + 5x_2 = x_0$
subject to: $3x_1 + 4x_2 \leq 14$
$$4x_1 + 2x_2 \leq 8$$
$$2x_1 + x_2 \leq 6.$$

The final tableau is:

	x_1	x_2	x_3	x_4	x_5	r.h.s.
	0	1	$\frac{2}{5}$	$-\frac{3}{10}$	0	$\frac{16}{5}$
	1	0	$-\frac{1}{5}$	$\frac{2}{5}$	0	$\frac{2}{5}$
	0	0	0	$-\frac{1}{2}$	1	2
x_0	0	0	$\frac{3}{5}$	$\frac{4}{5}$	0	$14\frac{4}{5}$

Suppose we change the r.h.s. constant of the first constraint from 14 to $14 + y$. Since x_3 is the slack variable for this constraint, all the r.h.s. values in the final tableau will change to

$$\frac{16}{5} + \frac{2}{5}y$$
$$\frac{2}{5} - \frac{1}{5}y.$$

However, in order that the solution is feasible these values must be non-negative. Thus

$$\frac{16}{5} + \frac{2}{5}y \geq 0 \Rightarrow y \geq -8$$
$$\frac{2}{5} - \frac{1}{5}y \geq 0 \Rightarrow y \leq 2.$$

Note that $y \geq -8$ implies that the r.h.s. constant must be greater than 6 and $y \leq 2$ implies that the r.h.s. constant must be smaller than 16 in order for the solution to be feasible. Thus the range is $-8 \leq y \leq 2$, with a r.h.s. constant range of 6 to 16. This means that for the problem to have an optimal and feasible solution the number of boxes of grapes can be no less than 6 or no greater than 16.

(d) In the previous L.P. problem, one of the l.h.s. constraint coefficients is changed from its original value. It is possible to analyze the effect of this change by using sensitivity analysis rather than solving the entire problem again from scratch.

From part (b), the tableau:

x_1	x_2	x_3	x_4	x_5	r.h.s.
3	4	1	0	0	14
4	2	0	1	0	8
2	1	0	0	1	6
-5	-4	0	0	0	0

becomes at optimality:

x_1	x_2	x_3	x_4	x_5	r.h.s.
0	1	$\frac{2}{5}$	$-\frac{3}{10}$	0	$\frac{16}{5}$
1	0	$-\frac{1}{5}$	$\frac{2}{5}$	0	$\frac{2}{5}$
0	0	0	$-\frac{1}{2}$	1	2
0	0	$\frac{3}{5}$	$\frac{4}{5}$	0	$\frac{74}{5}$

If a_{31} becomes $7\frac{1}{2}$ instead of 2, the same iterations produce:

x_1	x_2	x_3	x_4	x_5	r.h.s.
0	1	$\frac{2}{5}$	$-\frac{3}{10}$	0	$\frac{16}{5}$
1	0	$-\frac{1}{5}$	$\frac{2}{5}$	0	$\frac{2}{5}$
$\frac{11}{2}$	0	0	$-\frac{1}{2}$	1	2
0	0	$\frac{3}{5}$	$\frac{4}{5}$	0	$\frac{74}{5}$

which in canonical form is

x_1	x_2	x_3	x_4	x_5	r.h.s.
0	1	$\frac{2}{5}$	$-\frac{3}{10}$	0	$\frac{16}{5}$
1	0	$-\frac{1}{5}$	$\frac{2}{5}$	0	$\frac{2}{5}$
0	0	$\frac{11}{10}$	$-\frac{27}{10}$	1	$-\frac{1}{5}$
0	0	$\frac{3}{5}$	$\frac{4}{5}$	0	$\frac{74}{5}$

This is not feasible, as $x_3 < 0$. Using the dual simplex method, x_4 replaces x_5 in the basis (the only negative ratio):

x_1	x_2	x_3	x_4	x_5	r.h.s.
0	1	$\frac{5}{18}$	0	$-\frac{1}{9}$	$\frac{29}{9}$
1	0	$-\frac{7}{27}$	0	$\frac{4}{27}$	$\frac{10}{27}$
0	0	$-\frac{11}{27}$	1	$-\frac{10}{27}$	$\frac{2}{27}$
0	0	$\frac{25}{27}$	0	$-\frac{8}{27}$	$\frac{398}{27}$

Thus the new optimal solution is

$$x_1^* = \frac{10}{27},$$

$$x_2^* = \frac{29}{9},$$

$$x_0^* = \frac{398}{27}.$$

5. Model Validation

The primal and dual problems formulated in Section 3 have been validated by
LINDO and MAPLE, respectively (see Figure A5.1 and Figure A5.2).

```
MAXIMIZE 5x1 + 4x2
SUBJECT TO
        3x1 + 4x2 <= 14
        4x1 + 2x2 <= 8
        2x1 + x2 <= 6
        x1 >= 0
        x2 >= 0
END
LP OPTIMUM FOUND AT STEP        2
        OBJECTIVE FUNCTION VALUE
        1)      14.80000
   VARIABLE         VALUE          REDUCED COST
        X1          0.400000          0.000000
        X2          3.200000          0.000000
        ROW     SLACK OR SURPLUS     DUAL PRICES
        2)          0.000000          0.600000
        3)          0.000000          0.800000
        4)          2.000000          0.000000
        5)          0.400000          0.000000
        6)          3.200000          0.000000
  NO. ITERATIONS=        2
```

Figure A5.1a. Optimal solution to the wine production problem (primal) using the linear programming
package LINDO.

```
MINIMIZE 14 y1 + 8y2 + 6y3
SUBJECT TO
        3y1 + 4y2 + 2y3 >= 5
        4y1 + 2y2 + y3 >= 4
        y1 >= 0
        y2 >= 0
        y3 >= 0
END
LP OPTIMUM FOUND AT STEP          1
          OBJECTIVE FUNCTION VALUE
        1)          14.80000
   VARIABLE            VALUE            REDUCED COST
      Y1              0.600000            0.000000
      Y2              0.800000            0.000000
      Y3              0.000000            2.000000
      ROW     SLACK OR SURPLUS        DUAL PRICES
      2)              0.000000           -0.400000
      3)              0.000000           -3.200000
      4)              0.600000            0.000000
      5)              0.800000            0.000000
      6)              0.000000            0.000000
NO. ITERATIONS=          1
```

Figure A5.1b. Optimal solution to the wine production problem (dual) using the linear programming package LINDO.

```
> with(simplex):
  cnsts := {3*x1+4*x2<=14,4*x1+2*x2<=8,2*x1+x2<=6}:
  obj := 5*x1+4*x2:
  maximize(obj,cnsts union {x1>=0,x2>=0});
```
$$\{x2 = \frac{16}{5}, x1 = \frac{2}{5}\}$$

Figure A5.2a. Optimal solution to the wine production problem (primal) using the mathematical software package MAPLE.

```
> with(simplex):
  cnsts := {3*y1+4*y2+2*y3>=5,4*y1+2*y2+y3>=4}:
  obj := 14*y1+8*y2+6*y3:
  minimize(obj,cnsts union {y1>=0,y2>=0,y3>=0});
```
$$\{y3 = 0, y2 = \frac{4}{5}, y1 = \frac{3}{5}\}$$

Figure A5.2b. Optimal solution to the wine production problem (dual) using the mathematical software package MAPLE.

The results obtained in Sections 4.1 (b) and (c) have also been validated using the linear programming package LINDO (see Figure A5.3)

```
RANGES IN WHICH THE BASIS IS UNCHANGED:
                              OBJ COEFFICIENT RANGES
    VARIABLE          CURRENT        ALLOWABLE          ALLOWABLE
```

	COEF	INCREASE	DECREASE
X1	5.000000	3.000000	2.000000
X2	4.000000	2.666667	1.500000

RIGHTHAND SIDE RANGES

ROW	CURRENT RHS	ALLOWABLE INCREASE	ALLOWABLE DECREASE
2	14.000000	2.000000	8.000000
3	8.000000	4.000000	1.000000
4	6.000000	INFINITY	2.000000
5	0.000000	0.400000	INFINITY
6	0.000000	3.200000	INFINITY

Figure A5.3. Sensitivity analysis of Sections 4 (b) and (c) using the linear programming package LINDO.

6. Interpretation and Conclusions

This case study formulates a wine production problem in the form of a linear programming problem. It is intended to be illustrative in that it involves the use of the simplex method (by hand) in finding the optimal solution in the primal problem by interpreting the optimal dual tableau. Furthermore, sensitivity analysis is demonstrated by making small amendments to the key parameters. For problems of this type, model validation is essential. In fact, most realistic problems may be so complex that the computer is the only option. So it is important to have some experience in the use of appropriate L.P. software packages (e.g., LINDO, EXCEL, etc.) and popular computer algebra systems (e.g., MAPLE, MATHEMATICA, etc.).

7. Computer Algorithms

As this case study has been formulated in the form of a L.P. problem, it has been possible to validate the results using both the L.P. software package LINDO and the mathematical software package MAPLE.

8. References and Bibliography

1. Bunday, B.D., *Basic Optimization Methods*, Edward Arnold, London, 1984.
2. Cooper, L. and Sternberg, D., *Introduction to Methods of Optimization*, W.B. Saunders, Philadelphia, 1970.
3. Dantzig, G.B., *Linear Programming and Extensions*, Princeton University Press, Princeton, N.J., 1963.
4. Meerschaert, M.M., *Mathematical Modelling*, 2nd ed., Academic Press, London, 1999.
5. Press, W., Flannery, B., Teukolsky, S. and Vetterling, W. *Numerical Recipes*, Cambridge University Press, New York, 1987.
6. Sherali, H.D., Jarvis, J.J. and Bazaraa, M.S., *Linear Programming and Network Flows*, 2nd ed., Wiley, New York, 1990.
7. Ziegler, M.R. and Barnett, R.A., *Applied Mathematics for Business, Economics, Life Sciences and Social Sciences*, 6th ed., Prentice Hall, New York, 1996.

Case Study A6

TRANSPORTATION PROBLEM INVOLVING BREWERIES AND HOTELS

SUMMARY: In the business world, a Manager must recognize the typical task of allocating units from sources of supply to destinations of demand to minimize cost and that various transportation methods can be applied to effect this allocation of units. How to distribute products in such a manner as to minimize the total cost of their distribution constitutes a good example of an everyday problem that transportation methods can be used to solve.

This case study involves the supply system of 4 breweries, which supply the needs of 4 hotels. Figures are available for the production capacities (barrels/day) of the breweries and the demands (barrels/day) of the hotels. The transportation costs for a barrel of beer from each brewery to each hotel are also tabulated. A transportation problem is formulated in mathematical form and solved to produce a minimum cost schedule by using a number of approaches, including:

(a) northwest corner method;
(b) least cost method;
(c) Vogel approximation method;
(d) stepping stone algorithm and Dantzig's method.

The results are validated by comparing results from the various methods and by using the linear programming package LINDO and the mathematical software package MAPLE.

1. Background

The transportation problem is a special type of linear programming. It can be solved more efficiently by a modification of the simplex method than by the simplex method itself because of its structure. This can be demonstrated by considering a supply system comprising three factories which must supply the needs for a single commodity of three warehouses. The unit cost of shipping one item from each factory to each warehouse is known. There is a known limit to the production capacity of each factory. Also, each warehouse must receive a minimum number of units of the commodity. The problem is to find the minimum cost supply schedule which satisfies the production and demand constraints.

Problems which belong to this class of L.P. problems are called "transportation problems". However, many of the problems of this class do not involve the transportation of a commodity between sources and destinations. The algorithm for the solution of the transportation problem assumes that supply and demand are balanced. Of course, well formulated problems may arise in which "supply" exceeds "demand", or vice versa, as the problems may not involve the transportation of a commodity. In such cases, a fictitious "warehouse" or "factory" is introduced, whichever is required. Its "capacity" or "demand" is defined so as to balance total supply with total demand. All unit transportation costs to or from the fictitious location are defined to be zero. Then the value of the optimal solution to this balanced problem equals that of the original problem.

As already mentioned, because of its structure, the transportation problem can be solved efficiently by a modified simplex procedure. This structure is as follows:

(1) All l.h.s. constraint coefficients are either zero or one.
(2) All l.h.s. unit coefficients are always positioned in a distinctive pattern in the initial simplex tableau representing the problem (ignoring slack variables).
(3) All r.h.s. constraint constants are integers.

This structure implies that the optimal values of the decision variables will be integer, which is a very important result.

In solving problems by hand using the simplex method, it is convenient to display each iteration in a tableau. This is also done in the transportation problem, except a different type of tableau is used. The difference is that the value of each decision variable is written in each cell. There are a number of important methods by which an initial feasible solution can be identified and these are outlined below.

1.1 THE NORTHWEST CORNER METHOD

This method starts by allocating as much as possible to the cell in the northwest corner of the tableau of the problem, i.e., cell (1,1) (row 1, column 1). This means that we look in the row and column margins for the smallest demand or supply applicable and write the number in the northwest corner cell. Further cell allocations are made by moving down or to the right in the direction of leftover demand or supply and inserting the maximum feasible quantity at each step. We stop when the southeast corner has been allocated.

1.2 THE LEAST COST METHOD

Although the northwest corner method is easy to implement and always produces a feasible solution, it takes no account of the relative unit transportation costs. It is quite likely that the solution thus produced will be far from optimal. The methods of this section and the next usually produce less costly initial solutions. The least cost method starts by allocating the largest possible amount to the cell in the tableau with the least unit cost. This procedure will always satisfy a row or column which is removed from consideration. The cell with the next smallest unit cost is identified and the maximum is allocated to it. This procedure continues until all demand is met.

1.3 THE VOGEL APPROXIMATION METHOD

The Vogel approximation method often produces solutions which are even better than those of the least cost method. However, the price of this attractiveness is considerably more computation than the previous two methods. The Vogel approximation method begins by first reducing the matrix of unit costs. This reduction is achieved by subtracting the minimum quantity in each row from all elements in that row. The costs are further reduced by employing this procedure on the columns of the new cost matrix.

A penalty is then calculated for each cell which currently has zero unit cost. Each cell penalty represents the unit cost incurred if a positive allocation is not made to that cell. Each cell penalty is found by adding together the second smallest costs of the row and column of the cell.

The cell with the largest penalty is identified. The maximum amount possible is then allocated to this cell. Ties are settled arbitrarily. This procedure will always satisfy a row or column (or both), which is then removed from further consideration. This removal may require a further reduction in the cost matrix and a recalculation of some penalties. This process is repeated until all demand is met.

1.4 THE STEPPING STONE ALGORITHM

Once an initial feasible solution has been found by one of the three preceding methods, it can be transformed into the optimal solution. This is achieved by the stepping stone algorithm. To determine whether an initial feasible solution is optimal or not, it is necessary to ask, for each cell individually, if the allocation of one unit to that cell would reduce the total cost. This is done only for those cells which presently have no units assigned to them.

1.5 DANTZIG'S METHOD

The stepping stone method guarantees to find the minimal solution for any well formulated transportation problem in a finite number of steps. However, its implementation becomes very laborious on all but the smallest problems. For realistically sized problems a simpler method due to Dantzig is recommended. Like the stepping stone method it evaluates each empty cell in order to decide whether it would be profitable to make a positive assignment to it. This evaluation is based on the theory of duality, i.e., values are calculated for variables in the dual of the transportation problem regarded as a L.P. problem.

Unlike the stepping stone method, Dantzig's method does not create a circuit of cells in order to evaluate the value of an empty cell. Instead, it calculates values for the dual variables and these enable one to determine which empty cell should be filled. It then creates one circuit of cells in order to determine how much should be allocated and which cell leaves the basis. As only one circuit is created at each iteration, this method is much simpler than the preceding one.

1.6 CASE STUDY

The above methods are tested out on a transportation problem involving the supply system of four breweries, supplying the needs of four hotels for beer. Transportation costs are available for a barrel of beer from each brewery to each hotel. The production capacities (barrels/day) of the breweries and the demands (barrels/day) of the hotels are also given. The methods outlined in the previous sections are employed to help find the minimum cost schedule and the results are validated by comparing results from the various methods. A modification is included to show clearly the steps involved in using the stepping stone algorithm and Dantzig's method. Further validation is achieved by using LINDO and MAPLE.

2. Problem Statement

We consider the following transportation problem which involves the supply system of 4 breweries, supplying the needs of 4 hotels for beer. The transportation cost for a barrel of beer from each brewery to each hotel is given in Table A6.1. The production capacities of breweries 1, 2, 3 and 4 are 20, 10, 10 and 5 barrels per day, respectively. The demands of hotels A, B, C and D are 5, 20, 10 and 10 barrels per day, respectively. Find the minimum cost schedule.

		Hotels			
		A	B	C	D
	1	8	14	12	17
Breweries	2	11	9	15	13
	3	12	19	10	6
	4	12	5	13	8

Table A6.1. Transportation costs for a barrel of beer from each brewery to each hotel.

We will find an initial basis using the

(a) northwest corner method;
(b) least cost method;
(c) Vogel approximation method.

For model validation purposes we will solve the problem by the stepping stone algorithm and by Dantzig's method, starting with each basis. Further validation will be obtained using LINDO and MAPLE.

Finally, we will solve the problem again by changing both the production capacity of brewery 4 and the demand of hotel A from 5 to 15 barrels per day.

3. Model Formulation

Production constraints for breweries 1, 2, 3, 4 are

$$x_{11} + x_{12} + x_{13} + x_{14} \leq 20$$
$$x_{21} + x_{22} + x_{23} + x_{24} \leq 10$$
$$x_{31} + x_{32} + x_{33} + x_{34} \leq 10$$
$$x_{41} + x_{42} + x_{43} + x_{44} \leq 5.$$

Demand constraints for hotels A, B, C, D are

$$x_{11} + x_{21} + x_{31} + x_{41} \geq 5$$
$$x_{12} + x_{22} + x_{32} + x_{42} \geq 20$$
$$x_{13} + x_{23} + x_{33} + x_{43} \geq 10$$
$$x_{14} + x_{24} + x_{34} + x_{44} \geq 10.$$

All quantities transported must be nonnegative. Thus

$$x_{ij} \geq 0, \; i, j = 1, 2, 3, 4.$$

The objective is to find a supply schedule with minimum cost. The total cost is the sum of all costs from all breweries to all hotels. This cost x_0 can be expressed as

$$x_0 = 8x_{11} + 14x_{12} + 12x_{13} + 17x_{14} + 11x_{21} + 9x_{22} + 15x_{23} + 13x_{24}$$
$$+ 12x_{31} + 19x_{32} + 10x_{33} + 6x_{34} + 12x_{41} + 5x_{42} + 13x_{43} + 8x_{44}.$$

The problem can now be summarized in linear programming form:

Minimize: $x_0 = 8x_{11} + 14x_{12} + 12x_{13} + 17x_{14} + 11x_{21} + 9x_{22} + 15x_{23} + 13x_{24}$
$$+ 12x_{31} + 19x_{32} + 10x_{33} + 6x_{34} + 12x_{41} + 5x_{42} + 13x_{43} + 8x_{44},$$

subject to: $x_{11} + x_{12} + x_{13} + x_{14} \leq 20$
$$x_{21} + x_{22} + x_{23} + x_{24} \leq 10$$
$$x_{31} + x_{32} + x_{33} + x_{34} \leq 10$$
$$x_{41} + x_{42} + x_{43} + x_{44} \leq 5$$
$$x_{11} + x_{21} + x_{31} + x_{41} \geq 5$$
$$x_{12} + x_{22} + x_{32} + x_{42} \geq 20$$
$$x_{13} + x_{23} + x_{33} + x_{43} \geq 10$$
$$x_{14} + x_{24} + x_{34} + x_{44} \geq 10$$
$$x_{ij} \geq 0, i, j = 1, 2, 3, 4.$$

The tableau for this case study is given below.

Breweries		A	B	C	D	Supply
	1	8	14	12	17	20
	2	11	9	15	13	10
	3	12	19	10	6	10 (Hotels)
	4	12	5	13	8	5
Demand		5	20	10	10	

4. Mathematical/Numerical Solution

Identification of initial feasible solution by the four required methods follows.

4.1 THE NORTHWEST CORNER METHOD

The method starts by allocating as much as possible to the cell in the northwest corner of the tableau of the problem, cell (1,1) or row 1, column 1. The maximum that can be allocated is 5 units, as the demand of hotel A is 5 units. Column 1 is removed and cell (1,2) becomes the new northwest corner. A maximum of 15 units is allocated to this cell, all that remains in brewery 1. Row 1 is removed and cell (2,3) becomes the new northwest corner. This procedure continues until all demand is met. The tableau shows the feasible solution obtained.

Breweries	A	B	C	D	Supply
1	8 — 5	14 — 15	12	17	20
2	11	9 — 5	15 — 5	13	10
3	12	19	10 — 5	6 — 5	10 (Hotels)
4	12	5	13	8 — 5	5
Demand	5	20	10	10	

Conclusion

Brewery 1 supplies 5 units to hotel A and 15 units to hotel B.
Brewery 2 supplies 5 units to hotel B and 5 units to hotel C.
Brewery 3 supplies 5 units to hotel C and 5 units to hotel D.
Brewery 4 supplies 5 units to hotel D.

Total cost: 490 units.

4.2 THE LEAST COST METHOD

This method starts by allocating the largest possible amount to the cell in the tableau with the least unit cost. This means allocating 5 units to cell (4,2), and row 4 is removed. The demand of cell (2,2) and cell (1,2) is reduced to 10 and 5 units respectively. The cell with the next smallest cost is identified, i.e., cell (3,4) and 10 units are allocated to it removing row 3 and column 4. This procedure continues until all the demand is met. The following tableau illustrates the feasible solution which is obtained.

Breweries		A	B	C	D	Supply	Hotels
	1	[8] 5	[14] 5	[12] 10	[17]	20	
	2	[11]	[9] 10	[15]	[13]	10	
	3	[12]	[19]	[10]	[6] 10	10	
	4	[12]	[5] 5	[13]	[8]	5	
Demand		5	20	10	10		

Conclusion

Brewery 1 supplies 5 units to hotel A, 5 units to hotel B and 10 units to hotel C.
Brewery 2 supplies 10 units to hotel B.
Brewery 3 supplies 10 units to hotel D.
Brewery 4 supplies 5 units to hotel B.

Total cost: 405 units.

4.3 THE VOGEL APPROXIMATION METHOD

This method begins by first reducing the matrix of unit costs. This reduction is achieved by subtracting the minimum quantity in each row from all elements in that row. This results in the following tableau:

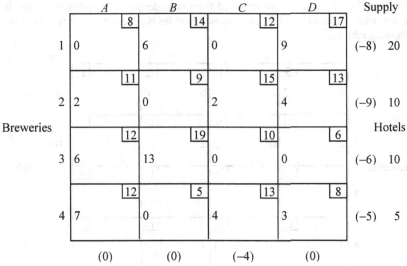

	A	B	C	D	Supply
1	[8] 0	[14] 6	[12] 4	[17] 9	(−8) 20
2	[11] 2	[9] 0	[15] 6	[13] 4	(−9) 10
3	[12] 6	[19] 13	[10] 4	[6] 0	(−6) 10
4	[12] 7	[5] 0	[13] 8	[8] 3	(−5) 5
Demand	5	20	10	10	

Breweries (left side), Hotels (right side).

The costs are further reduced by carrying out this procedure on the columns of the new cost matrix:

	A	B	C	D	Supply
1	[8] 0	[14] 6	[12] 0	[17] 9	(−8) 20
2	[11] 2	[9] 0	[15] 2	[13] 4	(−9) 10
3	[12] 6	[19] 13	[10] 0	[6] 0	(−6) 10
4	[12] 7	[5] 0	[13] 4	[8] 3	(−5) 5
	(0)	(0)	(−4)	(0)	

Breweries (left side), Hotels (right side).

A penalty is then calculated for each cell which currently has zero unit cost. Each cell penalty is found by adding together the second smallest costs of the row and column of the cell:

	A	B	C	D	Supply
1	[8] 0 2	[14] 6	[12] 0 0	[17] 9	(0) 20
2	[11] 2	[9] 0 2 2	[15] 4	[13]	(2) 10
3	[12] 6 13	[19] 0	[10] 0 0	[6] 3	(0) 10
4	[12] 7	[5] 0 3 4	[13] 3	[8]	(4) 5
	(2)	(0)	(0)	(4)	

Breweries (rows) — Hotels (columns)

The penalties are shown on the right-hand side of each appropriate cell and the cell with the largest penalty is identified. The maximum amount possible is then allocated to this cell. Cell (3,4) will be arbitrarily chosen and 10 units are allocated to it. Row 3 and column 4 are removed from consideration. A further reduction in the cost matrix and a recalculation of some penalties is necessary. This results in the following tableau:

	A	B	C	D	Supply	
1	[8] 0 2	[14] 6	[12] 0	[17] 0	20	(0)
2	[11] 2	[9] 0 2 2	[15]	[13]	10	(2)
3	[12]	[19]	[10]	[6] 10		
4	[12] 7	[5] 0 3 4	[13]	[8]	5	(4)
Demand	5	20	10			
	(2)	(0)	(2)			

Breweries (rows) — Hotels (columns)

Cell (4,2) is chosen and 5 units are allocated to it. Row 4 is then removed from consideration. This process is repeated until all demand is met.

	A	B	C	D	Supply
1	[8] 0	[14] 2 6	[12] 0	[17] 0	20 (0)
2	[11] 2	[9] 0 2 2	[15]	[13]	10 (2)
Breweries 3	[12]	[19]	[10]	[6] 10	Hotels
4	[12]	[5] 5	[13]	[8]	
Demand	5 (2)	20 (0)	10 (2)		

Cell (2,2) is chosen and 10 units are allocated to it, removing column 2 from consideration.

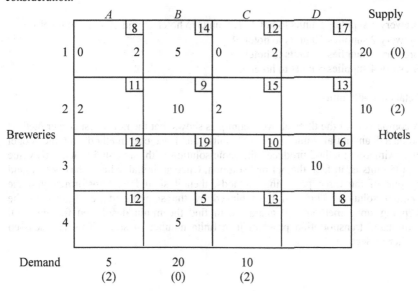

	A	B	C	D	Supply
1	[8] 0	[14] 2 5	[12] 0 2	[17]	20 (0)
2	[11] 2	[9] 10	[15] 2	[13]	10 (2)
Breweries 3	[12]	[19]	[10]	[6] 10	Hotels
4	[12]	[5] 5	[13]	[8]	
Demand	5 (2)	20 (0)	10 (2)		

Cell (1,1) is arbitrarily chosen and 5 units are allocated to it, removing column 1. Cell (1,3) must be allocated 10 units in order that all demand shall be met.

The final allocation is shown in the following tableau:

Breweries	A	B	C	D	Supply
1	[8] 5	[14] 5	[12] 10	[17]	20 (0)
2	[11]	[9] 10	[15]	[13]	10 (2)
3	[12]	[19]	[10]	[6] 10	10
4	[12]	[5] 5	[13]	[8]	5
Demand	5	20	10	10	

(Hotels)

Conclusion

Brewery 1 supplies 5 units to hotel A, 5 units to hotel B and 10 units to hotel C.
Brewery 2 supplies 10 units to hotel B.
Brewery 3 supplies 10 units to hotel D.
Brewery 4 supplies 5 units to hotel B.

Total cost: 405 units.

A comparison of the three above techniques shows that the northwest corner method produced an initial solution of 490 units, the least cost method and the Vogel approximation method produced the same solution with value 405 units. This value of 405 units is, in fact, the optimal solution. Once an initial solution has been found by one of the three preceding methods, then it should be transformed into the optimal solution. This can be achieved by the stepping stone algorithm. The stepping stone method will guarantee to find the minimal solution for any well formulated transportation problem in a finite number of steps. This is discussed further in Section 5.

5. Model Validation

5.1 STEPPING STONE ALGORITHM

Consider the initial feasible solution found by the northwest corner method. To determine whether this solution is optimal or not it is necessary to ask for each cell individually if the allocation of one unit to that cell would reduce the total cost. This is done for the cells which at present have no units assigned to them.

Cell (4,2) has the greatest decrease (11 units) and as much as possible, (5 units) is allocated to this cell. This means a decrease in cost of $(11 \times 5) = \$55$.

The new solution is displayed in the following tableau.

Breweries		A	B	C	D	Supply
	1	[8] 5	[14] 15	[12]	[17]	20
	2	[11]	[9]	[15] 10	[13]	10
	3	[12]	[19]	[10]	[6] 10	10
	4	[12]	[5] 5	[13]	[8]	5
Demand		5	20	10	10	

(Hotels: A, B, C, D)

The same procedure occurs – all empty cells in the new tableau are examined as before and the process is repeated. Since a basic feasible solution should contain $(m+n-1)$ basic variables, one of the empty cells is assigned a zero.

The process is repeated, but there is no allocation which will cause a cost reduction. Thus the optimal solution has already been found in Sections 4.2 and 4.3.

Conclusion

Brewery 1 supplies 5 units to hotel A, 5 units to hotel B and 10 units to hotel C.
Brewery 2 supplies 10 units to hotel B.
Brewery 3 supplies 10 units to hotel D.

Brewery 4 supplies 5 units to hotel B.

Total cost: 405 units.

It should be noted that the implementation of the stepping stone algorithm has not been continued to conclusion above as it is laborious to perform. This is true in general for all but the smallest problems. For realistically sized problems a simpler method due to Dantzig is recommended. This method is demonstrated in Section 5.3 for a modified problem obtained by altering some of the key parameters of the original problem.

5.2 MODIFIED PROBLEM

In order to clarify the steps involved in the stepping stone algorithm and Dantzig's method, the original problem is modified by changing both the production capacity of brewery 4 and the demand of hotel A from 5 to 15 barrels per day.

The modified tableau is now

	A	B	C	D	Supply
1	8	14	12	17	20
2	11	9	15	13	10
3	12	19	10	6	10
4	12	5	13	8	15
Demand	15	20	10	10	

Breweries — Hotels

First, we consider the initial feasible solution found by the northwest corner method.

	A	B	C	D	Supply
1	8 15	14 5	12	17	20
2	11	9 10	15	13	10
3	12	19 5	10 5	6	10
4	12	5	13 5	8 10	15
Demand	15	20	10	10	

Breweries (left)   Hotels (right)

Conclusion

Brewery 1 supplies 15 units to hotel A and 5 units to hotel B.
Brewery 2 supplies 10 units to hotel B.
Brewery 3 supplies 5 units to hotel B and 5 units to hotel C.
Brewery 4 supplies 5 units to hotel C and 10 units to hotel D.

Total cost: 570 units.

Next we use the stepping stone algorithm and Dantzig's method to determine whether this solution is optimal or not. We must ask for each cell individually if the allocation of one unit to that cell would reduce the total cost. This is done for the cells which at present have no units assigned to them.

Cell (4,2) has the greatest decrease (17 units) and as much as possible, (5 units) is allocated to this cell. This means a decrease in cost of $\$(17 \times 5) = \$85$.

The new solution is as follows:

Breweries	A	B	C	D	Supply
1	[8] 15	[14] 5	[12]	[17]	20
2	[11]	[9] 10	[15] 0	[13]	10
3	[12]	[19]	[10] 10	[6]	10 (Hotels)
4	[12]	[5] 5	[13]	[8] 10	15
Demand	15	20	10	10	

The same procedure occurs, i.e., all empty cells in the new tableau are examined as before and the process is repeated. Since a basic feasible solution should contain $(m+n-1)$ basic variables, one of the empty cells is assigned a zero. Cell (4,2) has the greatest decrease (19 units) and as much as possible (10 units) is allocated to this cell. This means a decrease in cost of $\$(19 \times 5) = \$95$. The new solution is then as follows:

Breweries	A	B	C	D	Supply
1	[8] 15	[14] 5	[12]	[17]	20
2	[11]	[9]	[15] 10	[13]	10
3	[12] 0	[19]	[10]	[6] 10	10 (Hotels)
4	[12]	[5] 15	[13]	[8]	15
Demand	15	20	10	10	

The process is repeated and cell (2,3) with a decrease of 8 units is allocated 5 units. This means a decrease in cost of $\$(8\times5)=\$40$. The tableau then becomes

Breweries		A	B	C	D	Supply	Hotels
	1	[8] 15	[14]	[12] 5	[17]	20	
	2	[11]	[9] 5	[15] 5	[13]	10	
	3	[12]	[19]	[10]	[6] 10	10	
	4	[12]	[5] 15	[13]	[8] 0	15	
Demand		15	20	10	10		

The process is repeated, but there is now no allocation which will cause a cost reduction. Thus the optimal solution has been found.

Conclusion

Brewery 1 supplies 15 units to hotel A and 5 units to hotel C.
Brewery 2 supplies 5 units to hotel B and 5 units to hotel C.
Brewery 3 supplies 10 units to hotel D.
Brewery 4 supplies 15 units to hotel B.

Total cost: 435 units.

5.3 COMPUTER VALIDATION

The results obtained in Sections 4.2, 4.3 and 4.4 have be validated by the linear programming package LINDO and MAPLE (see Figures A6.1 and A6.2).

```
MINIMIZE 8x11 + 14x12 + 12x13 + 17x14 + 11x21 + 9x22 + 15x23 + 13x24 +
12x31 + 19x32 + 10x33 + 6x34 + 12x41 + 5x42 + 13x43 + 8x44
SUBJECT TO
          x11 + x12 + x13 + x14 <= 20
          x21 + x22 + x23 + x24 <= 10
          x31 + x32 + x33 + x34 <= 10
          x41 + x42 + x43 + x44 <= 5
          x11 + x21 + x31 + x41 >= 5
          x12 + x22 + x32 + x42 >= 20
```

```
                    x13 + x23 + x33 + x43 >= 10
                    x14 + x24 + x34 + x44 >= 10
                    x11 >= 0
                    x12 >= 0
                    x13 >= 0
                    x14 >= 0
                    x21 >= 0
                    x22 >= 0
                    x23 >= 0
                    x24 >= 0
                    x31 >= 0
                    x32 >= 0
                    x33 >= 0
                    x34 >= 0
                    x41 >= 0
                    x42 >= 0
                    x43 >= 0
                    x44 >= 0
END
LP OPTIMUM FOUND AT STEP      11
        OBJECTIVE FUNCTION VALUE
        1)      405.0000
    VARIABLE            VALUE            REDUCED COST
         X11          5.000000            0.000000
         X12          5.000000            0.000000
         X13         10.000000            0.000000
         X14          0.000000            9.000000
         X21          0.000000            8.000000
         X22         10.000000            0.000000
         X23          0.000000            8.000000
         X24          0.000000            0.000000
         X31          0.000000            6.000000
         X32          0.000000            7.000000
         X33          0.000000            0.000000
         X34         10.000000            0.000000
         X41          0.000000           13.000000
         X42          5.000000            0.000000
         X43          0.000000           10.000000
         X44          0.000000            9.000000
        ROW      SLACK OR SURPLUS       DUAL PRICES
         2)          0.000000            0.000000
         3)          0.000000            5.000000
         4)          0.000000            2.000000
         5)          0.000000            9.000000
         6)          0.000000           -8.000000
         7)          0.000000          -14.000000
         8)          0.000000          -12.000000
         9)          0.000000           -8.000000
        10)          5.000000            0.000000
        11)          5.000000            0.000000
        12)         10.000000            0.000000
        13)          0.000000            0.000000
        14)          0.000000            0.000000
        15)         10.000000            0.000000
        16)          0.000000            0.000000
        17)          0.000000          -10.000000
        18)          0.000000            0.000000
```

19)	0.000000	0.000000
20)	0.000000	0.000000
21)	10.000000	0.000000
22)	0.000000	0.000000
23)	5.000000	0.000000
24)	0.000000	0.000000
25)	0.000000	0.000000

NO. ITERATIONS=   11

Figure A6.1. Optimal solution to the transportation problem using the linear programming package LINDO.

```
> with(simplex):
  cnsts := {x11+x12+x13+x14<=20,x21+x22+x23+x24<=10,
            x31+x32+x33+x34<=10,x41+x42+x43+x44<=5,
            x11+x21+x31+x41>=5,x12+x22+x32+x42>=20,
            x13+x23+x33+x43>=10,x14+x24+x34+x44>=10}:
  obj := 8*x11+14*x12+12*x13+17*x14+11*x21+9*x22+15*x23+13*x24+
         12*x31+19*x32+10*x33+6*x34+12*x41+5*x42+13*x43+8*x44:
  minimize(obj,cnsts union
  {x11>=0,x12>=0,x13>=0,x14>=0,x21>=0,x22>=0,x23>=0,x24>=0,
  x31>=0,x32>=0,x33>=0,x34>=0,x41>=0,x42>=0,x43>=0,x44>=0});
   {x21 = 0, x23 = 0, x24 = 0, x33 = 0, x43 = 0, x44 = 0, x31 = 0, x22 = 10, x13 = 10, x14 = 0,
    x32 = 0, x41 = 0, x34 = 10, x11 = 5, x42 = 5, x12 = 5}
```

Figure A6.2. Optimal solution to the transportation problem using the mathematical software package MAPLE.

6. Interpretation and Conclusions

This case study involves a typical transportation problem. Here the problem under consideration is the supply of beer from 4 breweries to 4 hotels. Rather than trying to use the simplex method, the problem can be solved more efficiently by using a modification of the simplex method. The case study has been carefully chosen so that the optimal values of the decision variables are integral. Three important techniques have been used to identify an initial feasible solution, namely,

(1) northwest corner method;
(2) least cost method;
(3) Vogel approximation method.

Once an initial feasible solution has been found by one of these methods, it is then important to transform it into the optimum solution. This has been achieved by using the stepping stone algorithm which guarantees to find the optimal solution for any well formulated transformation problem in a finite number of steps. This approach has been used in this case study for model validation purposes. Furthermore, the original problem has been modified and solved to help clarify the important steps in the solution process.

It should be noted that for realistically sized problems, the method due to Dantzig is recommended. Like the stepping stone method, it evaluates each empty cell in order to determine whether or not it should be profitable to make a positive assignment to it. Such evaluation is based on the theory of duality in L.P. problems. In effect, values are calculated for variables in the dual of the transportation problem which is regarded as a L.P. problem.

7. Computer Algorithms

The data for this particular case study have been particularly chosen so that the tableaux can be successfully manipulated by hand. In this way it has been possible to describe more clearly the various solution techniques. This has meant that computer algorithms have not been required. However, we do use LINDO and MAPLE to validate the results. Of course, for more complex larger problems, computer algorithms would be essential.

It is possible to solve large L.P. and transportation problems on a digital computer with the aid of properly organized calculations. In spite of the recent tremendous advancement in the computational power and memory size of modern computers, computational difficulties still arise in solving large L.P. problems. New techniques have been developed to overcome some of these. Such techniques include the revised simplex method, the dual simplex method, the primal-dual algorithm and Wolfe-Dantzig decomposition.

8. References and Bibliography

1. Lapin, L.L. and Whisler, W.D., *Quantitative Decision Making with the Spreadsheet Applications*, 7th ed., Duxbury, London, 2002.
2. Meredith, J.R., Shafer, S.M. and Turban, E., *Quantitative Business Modelling*, South-Western, Mason, Ohio, 2002.
3. Shore, B., *Quantitative Methods for Business Decisions: Text and Cases*, McGraw-Hill, New York, 1978.
4. Turban, E. and Meredith, J.R., *Fundamentals of Management Science*, 3rd ed., Business Publications Inc., 1985.

Project B7

PROFIT FROM AN ENGINEERING PLANT

SUMMARY: This project considers the manufacture of three major products (A , B and C) by an engineering plant. Figures are available on the resources required for the manufacture of each product together with the total resource availability. The problem is formulated as a linear programming (L.P.) problem which requires the maximization of the objective function (which is profit in this case) subject to three linear constraints which involve raw materials, production time and warehouse space. The L.P. problem is solved by hand using the simplex method and the results are validated using the L.P. software package LINDO. An extension is included to show the flexibility in the use of such software packages.

1. Background

This project deals with an optimization problem where constraints come into play and so it is classed as a constrained optimization problem. We are dealing with an engineering plant which manufactures three products and we have data on the raw materials, production time and profits for each product. This means that both the objective function and the constraints are linear. Hence we have a L.P. problem which involves maximizing profit subject to limited resources which take the form of known linear constraints.

Problems of this type can best be solved by the simplex method. In this case the problem is sufficiently simple to solve by hand using tableaux. However, it is important, for problems of this type, to validate the results using a L.P. software package. An advantage of such packages is that they are sufficiently user friendly to help answer other relevant questions. This may involve changes to the initial data, or in this case, examining the effects on the profit of making small increases on each of the resources. In this way it is possible to evaluate the most effective means of increasing the profits.

2. Problem Statement

An engineering plant makes three major products on a weekly basis. Each of these products requires a certain quantity of raw material and different production times, and yields different profits. The relevant information is displayed in Table B7.1. Note that there is sufficient warehouse space at the plant to store a total of 450kg/week.

(a) Set up a linear programming problem to maximize the profit.
(b) Solve the linear programming problem using the simplex method.
(c) Validate the results in (b) by solving the problem using a software package.
(d) Evaluate which of the following options will raise profits the most: increasing raw material, production time or storage.

Table B7.1. Data on the manufacture of products A, B and C by an engineering plant.

	Product A	Product B	Product C	Resource Availability
Raw material	5kg/kg	4kg/kg	10kg/kg	3000kg
Production time	0.05hr/kg	0.1hr/kg	0.2hr/kg	55hr/week
Profit	$30/kg	$30/kg	$35/kg	

3. Model Formulation

The objective of this problem is to maximize the profit of an engineering plant. The plant makes three major products, namely, A, B and C. Each product requires different amounts of raw material, has different production times and yields different profits. Furthermore, the problem is constrained because of the following limitations on resources:

(1) Total availabiltiy of raw materials = 3000kg.
(2) Total availabiltiy of production time = 55hr/week.
(3) Total availabiltiy of warehouse space at the plant = 450kg/week.

To formulate a mathematical model, we define the following list of variables:

x_1 = amount of Product A (kg)

x_2 = amount of Product B (kg)

x_3 = amount of Product C (kg).

We now deal in turn with the constraints on raw material, production time and warehouse space, respectively. The raw material required for Products A, B and C

is $5x_1$, $4x_2$ and $10x_3$, respectively. Since the available raw material is 3000kg/week, this leads to the constraint inequality

$$5x_1 + 4x_2 + 10x_3 \le 3,000. \tag{B7.1}$$

Similarly, the production time required for Products A, B and C is $0.05x_1$, $0.1x_2$ and $0.2x_3$, respectively. Since the available production time is 55hr/week, we have the second constraint inequality

$$0.05x_1 + 0.1x_2 + 0.2x_3 \le 55. \tag{B7.2}$$

Finally, since there is a limitation of 450kg/week on warehouse space, we have the third constraint inequality

$$x_1 + x_2 + x_3 \le 450. \tag{B7.3}$$

Our objective is to maximize the profit. Since the profit ($/kg) from Products A, B and C are 30, 30 and 35, respectively, the total profit, P, is given by

$$P = 30x_1 + 30x_2 + 35x_3. \tag{B7.4}$$

Letting $y = P$ be the quantity we wish to maximize with decision variables x_1, x_2 and x_3, then our problem is to maximize the objective function

$$y = f(x_1, x_2, x_3)$$
$$= 30x_1 + 30x_2 + 35x_3. \tag{B7.5}$$

This means that we have now formulated the following linear programming problem:

Maximize: $y = 30x_1 + 30x_2 + 35x_3$ (B7.6)

subject to: $5x_1 + 4x_2 + 10x_3 \le 3,000$

$0.05x_1 + 0.1x_2 + 0.2x_3 \le 55$ (B7.7)

$x_1 + x_2 + x_3 \le 450$

$x_1, x_2, x_3 \ge 0.$

In standard form the problem is

Maximize: $y = 30x_1 + 30x_2 + 35x_3$

subject to:
$$5x_1 \quad + 4x_2 \quad + 10x_3 \quad + x_4 \qquad\qquad\qquad = 3000$$
$$0.05x_1 \quad + 0.1x_2 \quad + 0.2x_3 \qquad\quad + x_5 \qquad = 55$$
$$x_1 \qquad + x_2 \qquad + x_3 \qquad\qquad\qquad + x_6 = 450$$
$$x_i \geq 0,\ i = 1, 2, \cdots, 6.$$

4. Mathematical Solution

This problem is now solved using the simplex method

	x_1	x_2	x_3	x_4	x_5	x_6	r.h.s.	Ratio
	5	4	10	1	0	0	3000	600
	0.05	0.1	0.2	0	1	0	55	1100
	$\boxed{1}$	1	1	0	0	1	450	450
y	−30	−30	−35	0	0	0	0	

	x_1	x_2	x_3	x_4	x_5	x_6	r.h.s.	Ratio
	0	−1	$\boxed{5}$	1	0	−5	750	150
	0	0.05	0.15	0	1	−0.05	32.5	217
	1	1	1	0	0	1	450	450
y	0	0	−5	0	0	30	13500	

	x_1	x_2	x_3	x_4	x_5	x_6	r.h.s.	Ratio
	0	−0.2	1	0.2	0	−1	150	
	0	$\boxed{0.08}$	0	−0.03	1	0.1	10	125
	1	1.2	0	−0.2	0	2	300	250
y	0	−1	0	1	0	−20	14250	

	x_1	x_2	x_3	x_4	x_5	x_6	r.h.s.
	0	0	1	0.125	2.5	−0.75	175
	0	1	0	−0.375	12.5	1.25	125
	1	0	0	0.25	−15	0.5	150
y	0	0	0	0.625	12.5	26.25	14375

The optimal solution is

$$x_1^* = 150,$$
$$x_2^* = 125,$$
$$x_3^* = 175,$$
$$y = 14375.$$

This means that a maximum profit of $14, 375 is obtained by producing 150kg of product A, 125kg of product B and 175kg of product C.

5. Model Validation

The results in Section 4 have been validated by using LINDO and MAPLE (see Figures B7.1 and B7.2, respectively).

```
MAXIMIZE 30x1 + 30x2 + 35x3
SUBJECT TO
        5x1 + 4x2 + 10x3 <= 3000
        0.05x1 + 0.1x2 + 0.2x3 <= 55
        x1 + x2 + x3 <= 450
        x1 >= 0
        x2 >= 0
        x3 >= 0
END
LP OPTIMUM FOUND AT STEP        3
        OBJECTIVE FUNCTION VALUE
        1)      14375.00
   VARIABLE         VALUE          REDUCED COST
        X1        150.000000         0.000000
        X2        125.000000         0.000000
        X3        175.000000         0.000000
        ROW    SLACK OR SURPLUS      DUAL PRICES
        2)         0.000000          0.625000
        3)         0.000000         12.500000
        4)         0.000000         26.250000
        5)       150.000000          0.000000
        6)       125.000000          0.000000
        7)       175.000000          0.000000
  NO. ITERATIONS=       3
```

Figure B7.1. Optimal solution to the engineering plant problem using the linear programming package LINDO.

```
> with(simplex):
  cnsts := {5*x1+4*x2+10*x3<=3000,
            0.05*x1+0.1*x2+0.2*x3<=55,
            x1+x2+x3<=450}:
  obj := 30*x1+30*x2+35*x3:
  maximize(obj,cnsts union {x1>=0,x2>=0,x3>=0};
            {x2 = 125., x3 = 175.0000000, x1 = 150.0000000}
```

Figure B7.2. Optimal solution to the engineering plant problem using the mathematical software package MAPLE.

In order to solve part (d) of the problem stated in Section 2, we consider separately the effects on the profit of increasing the raw material, production time and storage. Firstly, we increase the raw material by 10% and solve the L.P. problem again to determine the new maximum profit (see Figures B7.3a and B7.3b for the results using LINDO and MAPLE, respectively).

```
MAXIMIZE 30x1 + 30x2 + 35x3
SUBJECT TO
        5x1 + 4x2 + 10x3 <= 3300
        0.05x1 + 0.1x2 + 0.2x3 <= 55
        x1 + x2 + x3 <= 450
        x1 >= 0
        x2 >= 0
        x3 >= 0
END
 LP OPTIMUM FOUND AT STEP        3
        OBJECTIVE FUNCTION VALUE
        1)        14562.50
   VARIABLE          VALUE          REDUCED COST
        X1        225.000000          0.000000
        X2         12.500000          0.000000
        X3        212.500000          0.000000
        ROW    SLACK OR SURPLUS      DUAL PRICES
        2)          0.000000          0.625000
        3)          0.000000         12.500000
        4)          0.000000         26.250000
        5)        225.000000          0.000000
        6)         12.500000          0.000000
        7)        212.500000          0.000000
 NO. ITERATIONS=        3
```

Figure B7.3a. Optimal solution to the engineering plant problem with 10% increase in raw material using the linear programming package LINDO.

```
> with(simplex):
  cnsts := {5*x1+4*x2+10*x3<=3300,
            0.05*x1+0.1*x2+0.2*x3<=55,
            x1+x2+x3<=450}:
  obj := 30*x1+30*x2+35*x3:
  maximize(obj,cnsts union {x1>=0,x2>=0,x3>=0});
```
$$\{x2 = 12.50000000, x3 = 212.5000000, x1 = 225.0000000\}$$

Figure B7.3b. Optimal solution to the engineering plant problem with 10% increase in raw material using the mathematical software package MAPLE.

Next, we increase the production time by 10% and repeat the solution (see Figures B7.4a and B7.4b for the results using LINDO and MAPLE, respectively).

```
MAXIMIZE 30x1 + 30x2 + 35x3
SUBJECT TO
        5x1 + 4x2 + 10x3 <= 3000
        0.05x1 + 0.1x2 + 0.2x3 <= 60.5
        x1 + x2 + x3 <= 450
        x1 >= 0
        x2 >= 0
        x3 >= 0
END
 LP OPTIMUM FOUND AT STEP        3
        OBJECTIVE FUNCTION VALUE
        1)        14443.75
   VARIABLE          VALUE          REDUCED COST
        X1         67.500000          0.000000
```

```
        X2        193.750000          0.000000
        X3        188.750000          0.000000
       ROW    SLACK OR SURPLUS     DUAL PRICES
        2)         0.000000          0.625000
        3)         0.000000         12.500000
        4)         0.000000         26.250000
        5)        67.500000          0.000000
        6)       193.750000          0.000000
        7)       188.750000          0.000000
NO. ITERATIONS=       3
```

Figure B7.4a. Optimal solution to the engineering plant problem with 10% increase in production time using the linear programming package LINDO.

```
> with(simplex):
  cnsts := {5*x1+4*x2+10*x3<=3000,
            0.05*x1+0.1*x2+0.2*x3<=60.5,
            x1+x2+x3<=450}:
  obj := 30*x1+30*x2+35*x3:
  maximize(obj,cnsts union {x1>=0,x2>=0,x3>=0});
            {x3 = 188.7500000, x2 = 193.7500000, x1 = 67.50000000}
```

Figure B7.4b. Optimal solution to the engineering plant problem with 10% increase in production time using the mathematical software package MAPLE.

Finally, we increase the storage by 10% and repeat the solution (see Figures B7.5a and B7.5b for the results using LINDO and MAPLE, respectively).

```
MAXIMIZE 30x1 + 30x2 + 35x3
SUBJECT TO
        5x1 + 4x2 + 10x3 <= 3000
        0.05x1 + 0.1x2 + 0.2x3 <= 55
        x1 + x2 + x3 <= 495
        x1 >= 0
        x2 >= 0
        x3 >= 0
END
LP OPTIMUM FOUND AT STEP      3
          OBJECTIVE FUNCTION VALUE
        1)      15556.25
    VARIABLE        VALUE          REDUCED COST
        X1        172.500000          0.000000
        X2        181.250000          0.000000
        X3        141.250000          0.000000
       ROW    SLACK OR SURPLUS     DUAL PRICES
        2)         0.000000          0.625000
        3)         0.000000         12.500000
        4)         0.000000         26.250000
        5)       172.500000          0.000000
        6)       181.250000          0.000000
        7)       141.250000          0.000000
NO. ITERATIONS=       3
```

Figure B7.5a. Optimal solution to the engineering plant problem with 10% increase in storage using the linear programming package LINDO.

```
> with(simplex):
  cnsts := {5*x1+4*x2+10*x3<=3000,
            0.05*x1+0.1*x2+0.2*x3<=55,
            x1+x2+x3<=495}:
  obj := 30*x1+30*x2+35*x3:
  maximize(obj,cnsts union {x1>=0,x2>=0,x3>=0};
            {x1 = 172.5000000, x2 = 181.2500000, x3 = 141.2500000}
```

Figure B7.5b. Optimal solution to the engineering plant problem with 10% increase in storage using the mathematical software package MAPLE.

6. Interpretation and Conclusions

This engineering plant problem has been formulated as a L.P. problem and solved using the simplex method. The mathematical solution by hand using tableaux in Section 4 has been validated in Section 5 by using both LINDO and MAPLE. The extension of the problem, namely, part (d) in Section 2, has also been solved in Section 5. Profit figures have been obtained in Figures B7.3, B7.4 and B7.5 for a 10% increment in the three key parameters of raw material, production time and storage, respectively. These figures together with the respective percentage relative increase in profit figures are summarized in Table B7.2. From the figures in the final column, we can draw conclusions on the relative effectiveness of increasing the raw material, production time and storage. The respective profit ratio figures for raw material:production time:storage is 1.30:0.48:8.20 which implies 2.71:1:17.08. Clearly, storage space is the most important factor in terms of increasing profits.

Table B7.2. Increases in the profit due to 10% increments in the key parameters.

Parameter	Original maximum profit	Profit after 10% increment of the parameter	Relative increase of the profit (%)
Raw material	$14,375	$14,562.50	1.30%
Production time	$14,375	$14,443.75	0.48%
Storage	$14,375	$15,556.30	8.20%

A more detailed analysis has been carried out by performing separate simulations for further increases, namely, 20%, 30%, ..., 100% in the resource parameters. Consequently, profit ratio figures have been evaluated and these are presented in Table B7.3. Clearly, increasing storage will have the most effect on raising the profits. The maximum profit ratio of raw material: production time:storage is 1.67:1:49 and this occurs for the case 20%:20%:60% increase, respectively. Also it is clear that the profit figures stop increasing at certain percentage levels even after further increases in resources. In fact, there is no further increase in profits after 20%, 20% and 60% increase in raw materials, production time and storage, respectively.

Table B7.3. Table showing the effect on the profit of different percentage increments in each of the resources.

	Raw material	Production time	Storage	Profit ratio
Profit of plant after 10% increase of	$14,562.50	$14,443.75	$15,556.30	2.71:1:17.08
Profit of plant after 20% increase of	$14,583.33	$14,500.00	$16,737.50	1.67:1:18.89
Profit of plant after 30% increase of	$14,583.33	$14,500.00	$17,918.75	1.67:1:28.35
Profit of plant after 40% increase of	$14,583.33	$14,500.00	$19,100.00	1.67:1:37.08
Profit of plant after 50% increase of	$14,583.33	$14,500.00	$20,281.25	1.67:1:47.25
Profit of plant after 60% increase of	$14,583.33	$14,500.00	$20,500.00	1.67:1:49.00
Profit of plant after 70% increase of	$14,583.33	$14,500.00	$20,500.00	1.67:1:49.00
Profit of plant after 80% increase of	$14,583.33	$14,500.00	$20,500.00	1.67:1:49.00
Profit of plant after 90% increase of	$14,583.33	$14,500.00	$20,500.00	1.67:1:49.00
Profit of plant after 100% increase of	$14,583.33	$14,500.00	$20,500.00	1.67:1:49.00

7. Computer Algorithms

This project has been solved using a number of runs of the L.P. software package LINDO and the mathematical software package MAPLE.

8. References and Bibliography

1. Bunday, B.D., *Basic Optimization Methods*, Edward Arnold, London, 1984.
2. Cooper, L. and Sternberg, D., *Introduction to Methods of Optimization*, W.B. Saunders, Philadelphia, 1970.
3. Dantzig, G.B., *Linear Programming and Extensions*, Princeton University Press, Princeton, N.J., 1963.
4. Edwards, C., *Advanced Calculus of Several Variables*, Academic Press, New York, 1973.
5. Gill, P., Murray, W. and Wright, M., *Practical Optimization*, Academic Press, New York, 1981.
6. Polack, E., *Computational Methods in Optimization*, Academic Press, New York, 1971.

Project B8

OPTIMIZATION OF MANUFACTURE OF PERSONAL COMPUTERS

SUMMARY: In this project a manufacturer of personal computers is planning the introduction of two new products, a basic model and an enhanced model. Using unconstrained optimization which involves standard solution methods of multi-variable calculus, an analytical model is built to determine the production levels of the two types of computer. Constraints are then introduced based on the available production capacity and a model is developed using Lagrange multiplier methods. The problem is then reformulated as a linear programming problem by introducing some simplifying assumptions and solved to obtain the optimum production levels. Sensitivity analysis is also introduced involving both the profit and optimal production levels.

1. Background

Nowadays computers play a very important role in both the home and the workplace. Therefore the demands in the use of computers are great and different types of computers are being produced to meet the varying demands of people. Clearly, the formulation of a real life optimization problem involving computer manufacture will include a number of independent variables.

In this project a manufacturer of personal computers develops two new computer products. The total profits earned by the manufacturer will be affected by a number of factors, such as the number of computers sold, the selling price of the computers, the fixed costs of manufacturing, and so on. The objective here is to help the manufacturer decide how many computers should be made and sold in order to achieve the maximum annual profits.

In the case where there are no constraints (i.e., unconstrained optimization), a mathematical model can be developed and solved using the techniques of multi-variable calculus. However, this solution makes the assumption that the company has the potential to produce an unlimited number of computers per year. In a real life situation, constraints have to be imposed on the available production capacity which can depend on a number of factors. So a mathematical model can be formulated and solved using constrained optimization. This involves the use of Lagrange multiplier methods and linear programming to cater for the introduction of imposed constraints. Sensitivity analysis can also be employed to consider the effects of variations in the key parameters.

2. Problem Statement

A manufacturer of personal computers is planning the introduction of two new products, a basic model with retail price of $1250 and an enhanced model with retail price of $1500. The cost to the company is $700 and $850 for basic and enhanced models, respectively, plus an additional $500,000 in fixed costs. In a competitive market, the number of sales per year will affect the average selling price. It is estimated that for each type of computer, the average selling price will drop by $0.1 for each additional unit sold. Also, sales of the basic model will affect sales of the enhanced model, and vice versa. It is estimated that the average selling price for the basic model will be reduced by an additional $0.03 for each enhanced model sold, and the price for enhanced models will decrease by $0.04 for each basic model sold.

(a) Develop an analytical model to determine how many units of each type should be manufactured.

(b) Now introduce constraints based on the available production capacity. It is estimated that the available production capacity will be sufficient to produce 4,000 computers per year. Due to restriction of vital electronic parts, the supplier is able to supply parts for 2,000 basic models per year and for 3,000 enhanced models per year. How should the company set production levels?
[Suggestion: Develop a mathematical model using Lagrange multiplier methods]

(c) Reconsider the above mathematical model but now make the simplifying assumptions that the company makes a profit of $300 per basic model and $375 per enhanced model.
  (i) Find the optimum production levels by solving as a linear programming problem using the computer.
  (ii) Determine the shadow prices for each constraint and explain what they mean. Which constraints are binding on the optimal solution?
  (iii) Determine the sensitivity to the objective function coefficients (profit per unit). Consider both profit and optimal production levels.

(iv) Draw a graph of the feasible region and include a picture of ∇f at the optimum, where f is the objective function.

3. Model Formulation

To formulate our model, we first define a list of variables to be used, namely,

x_1 = number of basic model computers sold (per year)

x_2 = number of enhanced model computers sold (per year)

p_1 = selling price for basic model computers ($)

p_2 = selling price for enhanced model computers($)

C = cost of manufacturing computers ($/year)

R = revenue from the sale of computers ($/year)

P = profit from the sale of computers ($/year).

A list of assumptions is stated to guarantee the model is valid

$$p_1 = 1250 - 0.1x_1 - 0.03x_2$$
$$p_2 = 1500 - 0.04x_1 - 0.1x_2$$
$$R = p_1 x_1 + p_2 x_2$$
$$C = 500,000 + 700x_1 + 850x_2$$
$$P = R - C$$
$$x_1 \geq 0$$
$$x_2 \geq 0.$$

Our objective is to maximize the profit P ($/year) from the sale of computers.

The profit from the sale of the computers is given by

$$
\begin{aligned}
P &= R - C \\
&= p_1 x_1 + p_2 x_2 - 500,000 - 700x_1 - 850x_2 \\
&= (1250 - 0.1x_1 - 0.03x_2)x_1 + (1500 - 0.04x_1 - 0.1x_2)x_2 \\
&\quad -(500,000 + 700x_1 + 850x_2).
\end{aligned}
\qquad \text{(B8.1)}
$$

Letting $y = P$ be the quantity we wish to maximize with decision variables x_1 and x_2, then our problem is to maximize

$$y = f(x_1, x_2)$$
$$= (1250 - 0.1x_1 - 0.03x_2)x_1 + (1500 - 0.04x_1 - 0.1x_2)x_2 \quad \text{(B8.2)}$$
$$- (500{,}000 + 700x_1 + 850x_2),$$

for x_1 and x_2 nonnegative.

4. Mathematical Solution

4.1 UNCONSTRAINED OPTIMIZATION

Now, we proceed to solve the problem, using standard solution methods of the multi-variable calculus. The problem is to maximize the function given by (B8.2) over the region $x_1 \geq 0$, $x_2 \geq 0$. Figure B8.1 shows a 3-D graph of the function f.

We compute the partial derivatives of $f(x_1, x_2)$ with respect to x_1 and x_2 respectively, which, on setting equal to 0, give

$$\frac{\partial f}{\partial x_1} = 1250 - 0.2x_1 - 0.03x_2 - 0.04x_2 - 700 = 0,$$
$$\text{(B8.3)}$$
$$\frac{\partial f}{\partial x_2} = 1500 - 0.04x_1 - 0.2x_2 - 0.03x_2 - 850 = 0.$$

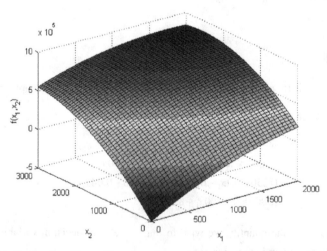

Figure B8.1. Graph of profit $y = f(x_1, x_2)$ versus production levels x_1 (basic computers) and x_2 (enhanced computers).

This is equivalent to solving the linear system

$$0.2x_1 + 0.07x_2 = 550,$$
$$0.07x_1 + 0.2x_2 = 650. \tag{B8.4}$$

which, using Cramer's rule, gives

$$x_1 = \begin{vmatrix} 550 & 0.07 \\ 650 & 0.2 \end{vmatrix} \Big/ D,$$

$$x_2 = \begin{vmatrix} 0.2 & 550 \\ 0.07 & 650 \end{vmatrix} \Big/ D, \tag{B8.5}$$

where D is the determinant of the coefficient matrix of equation (B8.4), which is given by $D = 0.2 \times 0.2 - 0.07 \times 0.07 = 0.0351$.

Therefore, we have

$$x_1 = \frac{110 - 45.5}{0.0351} = \frac{64.5}{0.0351} \cong 1{,}838,$$

$$x_2 = \frac{130 - 38.5}{0.0351} = \frac{91.5}{0.0351} \cong 2{,}607. \tag{B8.6}$$

The point (x_1, x_2) given by (B8.6) represents the global maximum of f. The maximum value of the objective function is obtained by substituting (B8.6) into equation (B8.2), which yields

$$y = (1250 - 0.1x_1 - 0.03x_2)x_1 + (1500 - 0.04x_1 - 0.1x_2)x_2$$
$$- (500{,}000 + 700x_1 + 850x_2)$$
$$\cong 852{,}564. \tag{B8.7}$$

The manufacturer simply needs to manufacture about 1838 units of the basic model and 2607 units of the enhanced model computers per year in order to achieve the maximum annual profit of $852,564. These figure indicate a profitable venture, so the company should be recommended to proceed with the introduction of these new products.

4.2 CONSTRAINED OPTIMIZATION

The objective function remains the same as in Section 3, namely,

$$y = f(x_1, x_2)$$
$$= (1250 - 0.1x_1 - 0.03x_2)x_1 + (1500 - 0.04x_1 - 0.1x_2)x_2$$
$$- (500{,}000 + 700x_1 + 850x_2),$$

subject to the constraints

$$x_1 \le 2{,}000$$
$$x_2 \le 3{,}000$$
$$x_1 + x_2 \le 4{,}000$$
$$x_1 \ge 0$$
$$x_2 \ge 0.$$

To find the maximum profit, we apply the method of Lagrange multipliers. We first compute

$$\nabla f = (550 - 0.2x_1 - 0.07x_2, 650 - 0.07x_1 - 0.2x_2). \qquad (B8.8)$$

Since $\nabla f \ne 0$ in the interior of the domain enclosed by the constraints, the maximum must occur on the boundary. Consider the segment of the boundary on the constraint line

$$g(x_1, x_2) = x_1 + x_2 = 4{,}000. \qquad (B8.9)$$

Hence, $\nabla g = (1,1)$, and so the Lagrange multiplier equations are

$$550 - 0.2x_1 - 0.07x_2 = \lambda$$
$$650 - 0.07x_1 - 0.2x_2 = \lambda. \qquad (B8.10)$$

Solving equations (B8.10) together with equation (B8.9), we have

$$x_1 = \frac{420}{0.26} \cong 1{,}615$$
$$x_2 \cong 2{,}385 \qquad (B8.11)$$
$$\lambda = 60.$$

Substituting the values in equation (B8.11) into the objective function, we have the estimated annual profit $\sim \$685{,}079$.

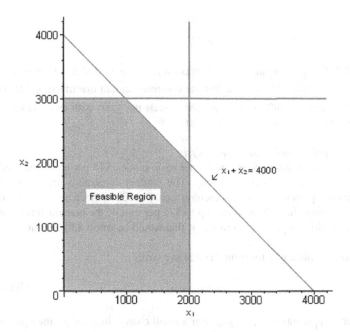

Figure B8.2. Graph showing the feasible region of production levels x_1 (basic computers) and x_2 (enhanced computers).

Figure B8.2 shows a graph of the feasible region for this problem. So the company can maximize profits by producing 1,615 basic computers and 2,385 enhanced computers for a total of 4,000 computers per year. This venture will produce an estimated profit of $685,079 per year.

4.3 COMPUTER SOLUTION (L.P. PROBLEM)

Now we assume the objective function

$$y = 300x_1 + 375x_2 .$$ (B8.12)

Our objective is to maximize $y = f(x_1, x_2) = 300x_1 + 375x_2$ subject to the constraints

$$x_1 \leq 2,000$$
$$x_2 \leq 3,000$$
$$x_1 + x_2 \leq 4,000$$

$$x_1 \geq 0$$
$$x_2 \geq 0.$$

We use LINDO to calculate that the maximum of y is 1,425,000 attained at $x_1 = 1,000$, $x_2 = 3,000$. We suggest that the company should manufacture 1,000 of the basic models and 3,000 of the enhanced models per year, with estimated profit of $1,425,000 per year (see Figure B8.5 later).

4.3.1 Shadow prices and senstitivity analysis

The shadow prices, or dual prices, are $0 per basic model, $75 per enhanced model and $300 per unit of production capacity. The binding constraints are those with non-zero shadow prices. If the manufacturer could produce more enhanced model computers, he should be willing to pay up to $75 per unit. If the manufacturer could free up some additional production capacity, that would be worth $300 per unit.

We generalize the objective function (B8.12) and write

$$y = c_1 x_1 + c_2 x_2, \tag{B8.13}$$

where currently $c_1 = 300$ and $c_2 = 375$. For a small change in c_1 or c_2, the optimum will still be at the corner point $(1000, 3000)$, and the sensitivity of x_1 or x_2 to c_1 or c_2 are all zero, that is,

$$S(x_1, c_1) = S(x_1, c_2) = 0$$
$$S(x_2, c_1) = S(x_2, c_2) = 0. \tag{B8.14}$$

The sensitivity of the objective function y to c_1 and c_2 is given by

$$S(y, c_1) = \frac{\partial y}{\partial c_1} \cdot \frac{c_1}{y} = (1,000) \frac{300}{1,425,000} \approx 0.21$$
$$S(y, c_2) = \frac{\partial y}{\partial c_2} \cdot \frac{c_2}{y} = (3,000) \frac{375}{1,425,000} \approx 0.79. \tag{B8.15}$$

Alternatively, we could use LINDO to re-optimize with the new objective functions

$$y = 303 x_1 + 375 x_2 \text{ and } y = 300 x_1 + 378.75 x_2. \tag{B8.16}$$

to see that x_1 and x_2 do not change, but y changes to

$$y = 1,428,000 \text{ and } y = 1,436,250, \tag{B8.17}$$

respectively.

We now consider the sensitivity of the optimal production levels x_1 and x_2 and the resulting profit y to the available manufacturing capacity of 4,000 units per year. We simply replace the equation (B8.9) by the more general form

$$g(x_1, x_2) = x_1 + x_2 = c. \tag{B8.18}$$

In this case, the feasible region is more or less the same to that pictured in Figure B8.2, but now the slanted constraint line is shifted a bit (with the same slope as $x_1 + x_2 = 4,000$ with x_2-intercept $= c$); see Figure B8.3. For values of c near 4,000, the maximum will be either $(2000, c - 2000)$ or $(c - 3000, 3000)$.

The level set with the largest value of c which still intersects the feasible region gives the maximum. Currently this occurs at $(1000, 3000)$.

Since $\nabla f = (300, 375) = (c_1, c_2)$, we consider the angle that the gradient vector (c_1, c_2) makes with the constraint line $x_1 + x_2 = 4,000$ at this point. This vector rotates as c_1 and c_2 change. We have the following 3 cases:

(i) $c_2 > c_1$, the situation is essentially the same as before.
(ii) $c_1 > c_2$, the optimal value moves to the adjacent corner point $(2000, c - 2000)$.
(iii) $c_1 = c_2$, the gradient vector is perpendicular to the constraint line $x_1 + x_2 = 4,000$ and so the level sets of (B8.13) are parallel to this line. In this case, the objective function attains its maximum at every point along the line segment joining the two corner points $(2000, c - 2000)$ and $(c - 3000, 3000)$; see Figure B8.4.

5. Model Validation

In order to make sure the results obtained in Section 4 are correct, we propose to use the mathematical software MAPLE and software package LINDO to validate our answers.

The optimal functional value of equation (B8.2) can be obtained using "maximize" command in MAPLE, that is

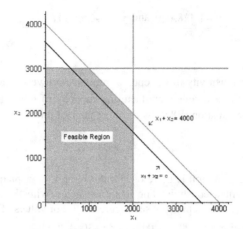

Figure B8.3. Graph showing the set of all feasible production levels x_1 of basic models and x_2 of enhanced models for the personal computer problem with constraints.

```
> maximize((1250-0.1*X1-0.03*X2)*X1+(1500-0.04*X1-0.1*X2)*X2-
  (500000+700*X1+850*X2),X1=0..infinity,X2=0..infinity,location);
      852564.100, {[{X1 = 1837.606838, X2 = 2606.837607}, 852564.100]}
```

Since both x_1 and x_2 are required to be integral, we choose $x_1 = 1838$, $x_2 = 2607$ as the number of computers to be manufactured in order to achieve the maximum profit of \$852,564.

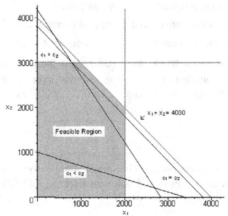

Figure B8.4. Graph showing the set of all feasible production levels x_1 of basic models and x_2 of enhanced models for the personal computer problem with constraints together with different gradient vector ∇f.

For the constrained optimization with linear coefficients, we use the optimization software to maximize the objective function $y = f(x_1, x_2) = 300x_1 + 375x_2$ (see Figure B8.5). Clearly, the company should manufacture 1,000 of the basic models and 3,000 of the enhanced models per year to give an estimated annual profit of $1,425,000.

```
MAXIMIZE 300X1 + 375X2
SUBJECT TO
           X1 <= 2000
           X2 <= 3000
           X1 + X2 <= 4000
           X1 >= 0
           X2 >= 0
END
LP OPTIMUM FOUND AT STEP      1
        OBJECTIVE FUNCTION VALUE
        1)      1425000.
    VARIABLE       VALUE          REDUCED COST
        X1      1000.000000        0.000000
        X2      3000.000000        0.000000
      ROW    SLACK OR SURPLUS     DUAL PRICES
        2)      1000.000000        0.000000
        3)         0.000000       75.000000
        4)         0.000000      300.000000
        5)      1000.000000        0.000000
        6)      3000.000000        0.000000
  NO. ITERATIONS=       1
```

Figure B8.5. Optimal solution to the personal computer problem using the linear programming package LINDO.

6. Interpretation and Conclusions

This project has considered the manufacture of two types of personal computer. Optimization was required in that advice must be given to the manufacturer to help him decide how many computers of each type should be made and sold to achieve maximal annual profits. This involved the optimization of a function of two variables in this case.

The project is illustrative in that it was first solved as an unconstrained optimization problem. Mathematical software such as MATLAB was helpful here particularly in producing 3D graphs. For nonlinear problems, computational techniques such as

Newton's method in several variables can prove helpful in locating the maximum or minimum point of the objective function.

Constraints were then added to demonstrate the techniques of constrained optimization. Here the method of Lagrange multipliers can play an important role. For simple projects such as this it was possible to use the computer to graph the feasible region. Furthermore, the results were validated by formulating as a L.P. problem and using a software package (in this case LINDO) to produce numerical results by means of the simplex method. Consequently, optimum production levels which maximize profits have been found. Information on the shadow prices for each constraint has been determined and explained. By generalizing the objective function, a sensitivity analysis has been performed which considers both profit and production levels.

It is important to note that such techniques can be extended to more realistic problems which involve higher dimensions and more complicated constraints. Clearly, computational techniques and the use of linear and non-linear programming become more important for such cases.

7. Computer Algorithms

In order to locate the optimal point of the objective function $f(x_1, x_2)$, we first compute $\nabla f(x_1, x_2)$ and solve for the zeros. However, as it is generally difficult to obtain the zeros analytically, we propose the use of Newton's method in two variables for such problems.

Algorithm (*Newton's method in two variables*)
Initial guess: x_1^0, x_2^0.
Define $F(x_1, x_2) = f_{x_1}(x_1, x_2)$ and $G(x_1, x_2) = f_{x_2}(x_1, x_2)$.
For $n = 1$ to N

$$a := F_{x_1}(x_1^{n-1}, x_2^{n-1})$$
$$b := F_{x_2}(x_1^{n-1}, x_2^{n-1})$$
$$c := G_{x_1}(x_1^{n-1}, x_2^{n-1})$$
$$d := G_{x_2}(x_1^{n-1}, x_2^{n-1})$$
$$p := -F(x_1^{n-1}, x_2^{n-1})$$
$$q := -G(x_1^{n-1}, x_2^{n-1})$$
$$D := ad - bc$$
$$x_1^n := x_1^{n-1} + (pd - qb)/D$$
$$y_1^n := y_1^{n-1} + (aq - cp)/D$$

END

For constrained optimization, the simplex method has been used.

8. References and Bibliography

1. Beltrami, E., *Models for Public Systems Analysis*, Academic Press, New York, 1977.
2. Courant, R., *Differential and Integral Calculus (Volume II)*, Wiley, New York, 1937.
3. Dantzig, G.B., *Linear Programming and Extensions*, Princeton University Press, Princeton, N.J., 1963.
4. Edwards, C., *Advanced Calculus of Several Variables*, Academic Press, New York, 1973.
5. Polack, E., *Computational Methods in Optimization*, Academic Press, New York, 1971.
6. Press, W., Flannery, B., Teukolsky, S. and Vetterling, W., *Numerical Recipes*, Cambridge University Press, New York, 1987.
7. Ziegler, M.R., and Barnett, R.A., *Applied Mathematics for Business Economics, Life Sciences and Social Sciences*, 6th ed., Prentice Hall, New York, 1996.

Project B9

AIR FREIGHT TRANSPORTATION PROBLEM

SUMMARY: This project involves the shipping of cargo by air. Apart from the weight constraints, the company has limited volume of aircraft storage compartments. Full details are available on an average daily basis of three types of cargo in terms of weight (tons) and volume (ft^3 / ton). The amount of each type of cargo which should be shipped by air each day is found in order to maximize revenue (\$). The problem is modelled using constrained optimization techniques and solved using Lagrange multipliers. The shadow prices are calculated for each constraint and the results are interpreted. An extension is considered which involves the company in reconfiguring some of its older planes to help increase the size of the cargo areas. In this way decisions can be made on whether or not to include alterations and to what extent. The numerical results are validated by solving as a linear programming problem using the computer software package LINDO and the mathematical software package MAPLE.

1. Background

Most optimization problems require the simultaneous consideration of a number of independent variables. The simplest category of multivariable optimization problems can be solved using unconstrained optimization and computer algebra systems such as MAPLE or MATHEMATICA prove helpful in dealing with the more complicated algebraic computations.

The simplest type of multivariable optimization problems involves finding the maximum or minimum of a differentiable function of several variables over a specified region. Complications arise in the solution of multivariable optimization models when the region over which we optimize is more complex. Most real problems lead to complicated models involving the existence of constraints on the independent variables. This leads us into the area of constrained optimization. One important technique for dealing with such problems involves the use of Lagrange multipliers. Essentially we have an objective function and a number of constraint functions. The set formed by the constraint functions is called the feasible region and the objective function is maximized or minimized over this set by the use of Lagrange multiplier techniques. In the solution process important information can be found by finding the shadow prices for each constraint.

The simplest type of multivariable constrained optimization problem is one where the objective function and the constraint functions are linear. The study of computational methods for such problems is called linear programming. Typical large-scale problems involve thousands of decision variables and thousands of constraints. The software for such problems is very flexible and user friendly and enables the user to amend the data and parameters as required which could produce results leading to important decision making.

2. Problem Statement

Consider a shipping company which uses air freight to move cargo which is stored in aircraft storage compartments. The company has the capacity to move 100 tons/day by air and charges $250/ton for air freight. Apart from the weight constraint, the company can only move 50,000ft^3 of cargo per day because of limited volume of aircraft storage compartments. Table B9.1 gives a breakdown of the average cargo available on a daily basis.

Table B9.1. Average cargo available each day.

Cargo	Volume (ft^3/ton)	Weight (tons)
A	550	30
B	800	40
C	400	50

(a) Formulate a mathematical model as a constrained optimization problem and solve using Lagrange multiplier techniques. Use this approach to determine how much of each cargo should be shipped by air each day in order to maximize revenue.

(b) Hence calculate the shadow prices for each constraint and explain the meaning of these results.

(c) Now consider the following extension of the problem. The company has the capacity to reconfigure some of its older aeroplanes to increase the size of the cargo areas. The alternations would cost $200,000 per aeroplane and would add 2,500ft^3 per aeroplane. The weight limits would be unchanged. Make the assumption that the aeroplanes fly 250 days per year and that the remaining lifetime of the older aeroplanes is approximately 5 years. Would it be worth while financially to make the alterations and, if so, to how many aeroplanes?

(d) Validate the above results by solving as a linear programming problem using an appropriate software package.

3. Model Formulation

To formulate our model, we first define a list of variables to be used, namely,

$$x_1 = \text{cargo } A \text{ (tons)}$$
$$x_2 = \text{cargo } B \text{ (tons)}$$
$$x_3 = \text{cargo } C \text{ (tons)}$$
$$V = \text{total volume (ft}^3)$$
$$W = \text{total weight (tons)}$$
$$F = \text{total freight charges (\$).}$$

The following assumptions are required to ensure the model is valid:

$$V = 550x_1 + 800x_2 + 400x_3$$
$$W = x_1 + x_2 + x_3$$
$$F = 250W$$
$$V \le 50,000$$
$$W \le 100$$
$$0 \le x_1 \le 30$$
$$0 \le x_2 \le 40$$
$$0 \le x_3 \le 50.$$

The objective of the model is to optimize the total freight charges F. Setting $y = F$, we can write

$$y = f(x_1, x_2, x_3) = 250x_1 + 250x_2 + 250x_3$$

and our objective is to optimize y over the domain for which

$$x_1 + x_2 + x_3 \leq 100$$
$$550x_1 + 800x_2 + 400x_3 \leq 50,000$$
$$0 \leq x_1 \leq 30$$
$$0 \leq x_2 \leq 40$$
$$0 \leq x_3 \leq 50.$$

4. Mathematical Solution

(a) We now solve the model using Lagrange multipliers. The objective function is linear and so the gradient of $f(x_1,x_2,x_3)$ is never zero. This means that there are no interior extreme points. The restriction of $f(x_1,x_2,x_3)$ to a plane or a line is still linear. This means that there are no local extrema along any of the faces or the edges of the feasible region. Hence the maximum must occur at one of the corners.

We now check each corner in turn to determine the optimum. Clearly it cannot be optimal to set any of x_1, x_2, x_3 equal to zero. Then there are five remaining linear constraints:

$$g_1(x_1,x_2,x_3) = x_1 + x_2 + x_3 = 100$$
$$g_2(x_1,x_2,x_3) = 550x_1 + 800x_2 + 400x_3 = 50,000$$
$$g_3(x_1,x_2,x_3) = x_1 = 30$$
$$g_4(x_1,x_2,x_3) = x_2 = 40$$
$$g_5(x_1,x_2,x_3) = x_3 = 50.$$

By solving any three of these we get the coordinates of a corner point. If these coordinates satisfy the inequality constraints, then the corner point represents a feasible solution. By checking all 10 corner points, we find that the optimum occurs at the intersection of constraint lines g_2, g_3 and g_5 given by $(x_1,x_2,x_3) =$ (30,16.875,50) at which we find

$$y = 250(x_1 + x_2 + x_3) = 24,218.75 .$$

So the optimal strategy is to ship the maximum of 30 tons/day of cargo A and 50 tons/day of cargo C. Then, because of volume constraints, we can only ship 16.875 tons/day of cargo B. This will yield a total shipping charge of \$24,218.75 per day. It is important to note that the weight constraint is not binding, i.e., we do not have enough volume in the cargo holds to ship all 100 tons of available cargo.

(b) Next we calculate and discuss the shadow prices for each constraint. The gradient vectors for the binding constraints are as follows:

$$v_2 = \nabla g_2 = (550, 800, 400)$$
$$v_3 = \nabla g_3 = (1, 0, 0)$$
$$v_5 = \nabla g_5 = (0, 0, 1)$$

and $w = \nabla f = (250, 250, 250)$ for the objective function f.

The Lagrange multiplier equations are

$$w = \lambda_2 v_2 + \lambda_3 v_3 + \lambda_5 v_5$$

so that

$$250 = 550\lambda_2 + \lambda_3$$
$$250 = 800\lambda_2$$
$$250 = 400\lambda_2 + \lambda_5.$$

Hence $\lambda_2 = 0.3125$, $\lambda_3 = 78.125$ and $\lambda_5 = 125$.

These figures give the shadow prices, which effectively mean that additional cargo capacity is worth approximately $0.31 per cubic foot. The net advantage of being able to ship more of cargo A is $78.13 per ton, and for cargo C the figure is $125 per ton.

For all of the other constraints, which are all nonbinding, the shadow prices are zero. So, for example, the company would not be willing to pay to increase the weight capacity of the aeroplanes, since the current optimal solution does not use all of the available weight capacity.

(c) Next we use the above sensitivity results to solve the extension to the problem. We know that additional cargo space is worth $0.31 per cubic foot. Over the useful lifetime of the aeroplanes (i.e., 5 years), the proposed modification would allow the company to ship

$$5 \times 250 \times 2000 = 2,500,000$$

additional cubic feet of cargo, which should be worth

$$\$0.31 \times 2,500,000 = \$775,000.$$

Since this figure is well above the $200,000 cost of the reconfiguration, the company should proceed with this plan.

The remaining cargo not currently being shipped consists of

$$40 - 16.875 = 23.125$$

tons per day of cargo B, which would fill a volume of

$$23.125 \times 800 = 18,500$$

cubic feet. The current daily load uses all of the available volume and weighs 96.875 tons. Each modified aeroplane can carry an additional 2,500 cubic feet of cargo B, which weighs

$$2,500/800 = 3.125 \text{tons}.$$

Although there is enough additional cargo to fill 7 or 8 modified aeroplanes, there is only enough total weighs capacity to carry an additional 3.125 tons per day. This means that we should only modify one of the aeroplanes. If we modify one plane then the binding constraints are g_3, g_5 and the modified constraint $g_2 \leq 52,500$.

So now we solve

$$550x_1 + 800x_2 + 400x_3 = 52,500$$

subject to

$$x_1 = 30, \ x_3 = 50.$$

This gives $x_2 = 20$ and so in this case we ship 20 tons of cargo B and obtain the shipping charges of $25,000 since

$$250(x_1 + x_2 + x_3) = 250(30 + 20 + 50) = 25,000.$$

5. Model Validation

Now we validate the results obtained in Section 4 by solving as a linear programming problem using the simplex method. We have chosen the LINDO software package to carry out the computation.

(a) Of course, the list of variables and assumptions are the same as in Section 4. So our objective is to maximize

$$250x_1 + 250x_2 + 250x_3$$

subject to the constraints

$$550x_1 + 800x_2 + 400x_3 \leq 50,000$$
$$x_1 + x_2 + x_3 \leq 100$$
$$x_1 \leq 30$$
$$x_2 \leq 40$$
$$x_3 \leq 50.$$

The L.P. optimum found at step 3 using LINDO is 24,218.75. This maximum is attained at $x_1 = 30$, $x_2 = 16.875$ $x_3 = 50$ (see Figure B9.1 for LINDO results and Figure B9.2 for MAPLE results). So we recommend that the company ships 30 tons of cargo A, 16.875 tons of cargo B and 50 tons of cargo C per day. This should result in the company collecting a shipping charge of $24,218.75 per day.

```
MAXIMIZE 250x1 + 250x2 + 250x3
SUBJECT TO
        550x1 + 800x2 + 400x3 <= 50000
        x1 + x2 + x3 <= 100
        x1 <= 30
        x2 <= 40
        x3 <= 50
        x1 >= 0
        x2 >= 0
        x3 >= 0
END
 LP OPTIMUM FOUND AT STEP        3
           OBJECTIVE FUNCTION VALUE
        1)      24218.75
    VARIABLE        VALUE          REDUCED COST
        X1        30.000000          0.000000
        X2        16.875000          0.000000
        X3        50.000000          0.000000
        ROW    SLACK OR SURPLUS      DUAL PRICES
        2)         0.000000          0.312500
        3)         3.125000          0.000000
        4)         0.000000         78.125000
        5)        23.125000          0.000000
        6)         0.000000        125.000000
 NO. ITERATIONS=       3
```

Figure B9.1. Optimal solution to the air freight transportation problem (a) using the linear programming package LINDO.

(b) Next we validate the shadow prices for each constraint found in Section 4. The Lagrange multipliers or shadow prices are now computed automatically by LINDO under the heading of "DUAL PRICES". Row 2 is the volume constraint and the corresponding dual price is 0.3125. This is the shadow price for volume.

```
> with(simplex):
  cnsts := {550*x1+800*x2+400*x3<=50000,
            x1+x2+x3<=100,
            x1<=30,
            x2<=40,
            x3<=50}:
  obj := 250*x1+250*x2+250*x3:
  maximize(obj,cnsts union {x1>=0,x2>=0,x3>=0});
```

$$\left\{ x1 = 30,\ x3 = 50,\ x2 = \frac{135}{8} \right\}$$

Figure B9.2. Optimal solution to the air freight transportation problem (a) using the mathematical software package MAPLE.

Row 4 is the upper bound on the amount of cargo A available, and the corresponding dual price is 78.125. This is the shadow price for a ton of cargo A. Row 6 is the upper bound on the amount of cargo C available, and the corresponding dual price is 125. This is the shadow price for a ton of cargo C. Extra cargo space is worth \$0.3125 per cubic foot. Additional cargo of type A is worth \$78.13 per ton, and additional cargo of type C is worth \$125 per ton.

(c) Finally, we validate the results for the extension to the original problem. The cost of upgrade is \$160 per aeroplane per day. To ascertain if it is better to modify one aeroplane, we modify the volume constraint so that it now becomes

$$550x_1 + 800x_2 + 400x_3 \leq 52,500$$

and reoptimize. The modification is worth while if the objective function increases by more than 160.

Using LINDO (see Figure B9.3) and MAPLE (see Figure B9.4) we find that that the new optimum is 25,000, so that the new net income is \$25,000 per day. This is clearly better than the result in (a) and so it is worth while modifying one plane. The optimum of 25,000 occurs at $x_1 = 30$, $x_2 = 20$ and $x_3 = 50$.

As there is a restriction of 100 tons on the total daily load, we should not modify more than one aeroplane.

The LINDO software package is very user friendly and can be run a number of times. For example, if we wished to decide whether it is better to modify a second aeroplane, we simply change the volume constraint to

$$550x_1 + 800x_2 + 400x_3 \leq 52,500$$

and reoptimize. Another LINDO run would yield the optimum and the corresponding values of x_1, x_2 and x_3. The net profit would be less than before. Note also that from here on we use all the available weight capacity and so there is no point in considering the addition of cargo capacity to any more aeroplanes.

```
MAXIMIZE 250x1 + 250x2 + 250x3
SUBJECT TO
         550x1 + 800x2 + 400x3 <= 52500
         x1 + x2 + x3 <= 100
         x1 <= 30
         x2 <= 40
         x3 <= 50
         x1 >= 0
         x2 >= 0
         x3 >= 0
END
LP OPTIMUM FOUND AT STEP       4
         OBJECTIVE FUNCTION VALUE
      1)       25000.00
   VARIABLE            VALUE           REDUCED COST
        X1         30.000000            0.000000
        X2         20.000000            0.000000
        X3         50.000000            0.000000
      ROW    SLACK OR SURPLUS          DUAL PRICES
      2)          0.000000             0.312500
      3)          0.000000             0.000000
      4)          0.000000            78.125000
      5)         20.000000             0.000000
      6)          0.000000           125.000000
  NO. ITERATIONS=        4
```

Figure B9.3. Optimal solution to the air freight transportation problem (c) using the linear programming package LINDO.

```
> with(simplex):
  cnsts := {550*x1+800*x2+400*x3<=52500,
            x1+x2+x3<=100,
            x1<=30,
            x2<=40,
            x3<=50}:
  obj := 250*x1+250*x2+250*x3:
  maximize(obj,cnsts union {x1>=0,x2>=0,x3>=0});
                    {x1 = 30, x3 = 50, x2 = 20}
```

Figure B9.4. Optimal solution to the air freight transportation problem (c) using the mathematical software package MAPLE.

6. Interpretation and Conclusions

This project is interesting in that it involves the use of air freight for transporting cargo by a shipping company. Since it only deals with three types of cargo and the key parameters of weight and volume, the number of constraints is small and therefore easily handled in a mathematical solution. Also the mathematical solution can easily be validated by the use of a number of linear programming software packages on the market.

It should be pointed out that more realistic shipping and transportation problems are usually much more complex because of the larger dimensions and the complexity of the constraints. So, for example, graphical techniques are not available for dimensions $n > 3$, and solving $\nabla f = 0$ becomes more complicated as the number of independent variables increases. Constrained optimization is also more difficult becomes the geometry of the feasible region can be more complicated.

In general, multivariable optimization problems with constraints are almost always difficult to solve. Of course, linear programming is an important tool in the solution of such problems where both the objective function and the constraint functions are linear. Software packages for linear programming are widely available and are in frequent use for problems in manufacturing, investment, transportation, farming and government. Typical large-scale problems involve thousands of decision variables and thousands of constraints.

Although a variety of computational techniques have been developed to handle special types of multivariable optimization problems, good general methods do not yet exist even at the most sophisticated levels. The area of research that covers the development of new computational methods for such problems is called nonlinear programming, and it is very active.

7. Computer Algorithms

The mathematical solution of this problem given in Section 4 has involved the use of MAPLE to obtain some numerical results. For the model validation in Section 5, the popular linear programming software package LINDO has been used to carry out a number of runs.

8. References and Bibliography

1. Beightler, C. Phillips, D. and Wilde, D., *Foundations of Optimization,* 2nd ed., Prentice-Hall, Englewood Cliffs, N.J., 1979.
2. Lapin, L.L., *Quantitative Methods for Business Decisions*, 6th ed., Duxbury Press, London, 1994.
3. Thierauf, R.J., Klekamp, R.C. and Ruwe, M.L., *Management Science: A Model Formulation Approach with Computer Applications*, Merrill Publishing Co., London, 1985.
4. Waters, D., *Quantitative Methods for Business*, 2nd ed., Addison-Wesley, New York, 1997.

Optimization Problems

1. Weekly production schedules are required for the manufacture of two products
   A and B. Each unit of A uses one component made in the factory, while each
   unit of B uses two of the components. The factory has a maximum output of
   100 components per week. Each unit of A and B requires 12 hours of
   subcontracted work and agreements have been arranged with subcontractors for
   a weekly minimum of 240 hours and a maximum of 600 hours. The marketing
   department says that all production of B can be sold but there is a maximum
   demand of 60 units of A, in spite of a long-term contract to supply 10 units of
   A to one customer. The net profit on each unit of A and B is $300 and $400,
   respectively.

   (a) Formulate the above as a linear programming problem.

   (b) Use a graphical method to find an optimum solution. Validate this result
   using the computer.

2. An electrical components manufacturer produces two types of tester, Basic and
   Enhanced. The production time (hours/hundred units) of each type and the
   capacity of each production process are given in Table 1 below. All testers made
   can be sold and the profits on each unit of Basic and Enhanced are $16 and $20,
   respectively.

   (a) By formulating as a linear programming problem, find the maximum
   monthly profit.

   (b) Find the spare capacities in production facilities. Also calculate the shadow prices of production facilities.

(c) It has been suggested that the selling price of the Basic tester should be raised. To what level could the profit be raised without changing the optimal production pattern?

(d) A new tester is planned, which could go through Pressing, Wiring and Assembly at a rate of 400 units/hour on each process. What profit is required before this new tester is made?

Table 1. Electrical components data.

Process	Basic	Enhanced	Capacity (hours/month)
Pressing	4	8	320
Wiring	12	4	480
Assembly	8	8	400

3. A box manufacturing company makes medium and extra-large units. Medium boxes each require 8 square feet of wood, while extra-large ones consume 12. All boxes require 1 hour of labour irrespective of size. Wood is limited to 480 square feet and only 24 hours of labour are available. Due to floor-space limitations, no more than 24 extra-large boxes can be made each day. Also, customers can absorb at most 30 medium boxes. Subject to these constraints, any number of medium boxes may be made. Medium boxes produce a profit of $6 but extra-large ones earn only $3.

Using the notation:

$$x_m = \text{number of medium boxes to be made}$$
$$x_l = \text{number of extra-large boxes to be made,}$$

show that this produces the linear programming problem:

Maximize: $P = 6x_m + 3x_l$

subject to:
$$8x_m + 12x_l \le 480 \quad \text{(wood)}$$
$$x_m + x_l \le 24 \quad \text{(labour)}$$
$$x_l \le 24 \quad \text{(floor space)}$$
$$x_m \le 30 \quad \text{(customer)}$$
$$x_m, \ x_l \ge 0.$$

By introducing slack variables, solve this problem using the simplex method. Validate the results by computer.

4. A food products company has established a new products division to develop and test market new snack foods. The manager of this division is considering three promising products: A, B and C. He feels that linear programming (the simplex method) offers the best means for determining an optimum production schedule which would allow producing these products simultaneously. The company has three basic manufacturing departments: mixing, frying and packing. The time requirements for each product and the total available monthly hours are as shown in Table 2 below.

Table 2. Time requirements for each product and total available monthly hours.

Product	Department		
	Mixing (hour)	Frying (hour)	Packing (hour)
A	0.2	0.2	0.1
B	0.3	0.5	0.1
C	0.4	0.3	0.1
Available monthly hours	6,000	5,000	4,000

It is estimated that the contribution for products A, B and C are $0.40, $0.50 and $0.60, respectively.

Use the simplex method to find the optimum quantity for each product and total contribution based on the data for the monthly time available in each department and for product contribution in Table 2. Validate the results by computer.

5. A cooker manufacturing company makes two types of cooker, one electric and one gas. There are four stages in the production of each of these, with details given in Table 3 below. The electric cooker has variable costs of $300 per unit and a selling price of $450 while the gas cooker has variable costs of $240 and a selling price of $360. Fixed overheads are $100,000 per week and the company works a 50-week year. The marketing department suggests maximum sales of 1000 electric and 1200 gas cookers per week.

(a) Formulate this as a profit-maximizing linear programming problem. Use the simplex method to find an optimal product mix for the company. What is the expected annual profit?

(b) Find the used and spare capacity of each manufacturing stage. What are the shadow prices of each manufacturing stage?

Table 3. Data on the production of electric and gas cookers.

Manufacturing stage	Time required (hours/unit)		Total time available (hours/week)
	electric	gas	
Forming	4	3	4,000
Machine shop	12	9	12,000
Assembly	6	3	6,000
Testing	2	2	3,000

(c) An outside consultant offers his testing services to the company. What price should the company be willing to pay for this service and how many hours should be bought per week?

(d) A new cooker is planned, which would use the manufacturing stages for 4, 10, 6 and 2 hours, respectively. At what selling price should the company consider making this cooker if the other variable costs are $250 per unit?

(e) Validate the above results, where possible, by computer.

6. A company has consistently followed a policy of producing only those products which contribute the highest amounts to fixed costs and profit. However, consideration has always been given to producing the minimum weekly sales requirements, which for products A, B, C and D are 20 units, 25 units, 25 units and 20 units, respectively. The production requirements and time available for next week are given in Table 4 below. At present the weekly production mix (considering the minimum sales requirements) is as follows:

product A, 1,500; product B, 30; product C, 40; product D, 30.

Table 4. Production requirements and time available for next week.

	Time required per product (hours)				Time available for next week (hours)
	A	B	C	D	
Department 1	0.2	0.25	0.1	0.2	500
Department 2	0.3	0.4	0.4	0.3	1,000
Department 3	0.25	0.3	0.2	0.25	500
Department 4	0.2	0.3	0.3	0.25	600
Contribution per unit	$10	$9	$8	$10	

Produce a mathematical model to help answer the following questions:

(a) Are the present product mix and contribution for the firm optimum? If not, what should they be?

(b) What recommendations concerning production facilities should be made to the firm?

Validate, where possible, the above answers by computer.

7. A company is considering manufacturing a new product line which includes four products. Each product can be made by two different and distinct methods, one of which has two processes and the other three. All products will be made on a second-shift basis. Data on the product's selling price and variable costs together with the probable quantities that can be sold on the basis of current marketing research are given in Table 5 below. Also the firm's manufacturing section has determined the manufacturing times for each process and this information is given in Table 6 below.

Table 5. The product's selling price and variable costs plus the probable quantities that can be sold on the basis of current marketing research.

	Product			
	1	2	3	4
Selling price to wholesaler (40% discount)	$100	$160	$130	$140
Variable costs – method A	$90	$140	$120	$135
Variable costs – method B	$110	$160	$100	$120
Quantity that can be sold	1,000	2,000	4,000	6,000

Table 6. Manufacturing times for each process.

	Product			
	1	2	3	4
Method A				
Department 10	2.8	3.4	2.0	3.6
Department 11	8.0	10.0	8.0	9.0
Department 12	1.0	0.8	0.5	0.5
Method B				
Department 21	4.0	3.0	2.0	4.0
Department 22	4.0	8.0	5.0	4.0

Monthly hours available are as follows:

Department 10	20,000
Department 11	40,000
Department 12	10,000
Department 21	10,000
Department 22	10,000.

Produce a mathematical model to decide what the firm should do, in light of the production times and the possible production bottlenecks, in order to maximize total monthly contribution.

8. An oil company makes two blends of fuel by mixing three oils. Figures on the costs and daily availability of the oils are given in Table 7 below. Also, the requirements of the blends of fuel are given in Table 8.

Table 7. Costs and daily availability of the oils.

Oil	Cost ($/litre)	Amount available (litres)
A	0.30	6,000
B	0.40	10,000
C	0.48	12,000

Table 8. Requirements of the blends of fuel.

Blend 1	at least 30% of A at most 50% of B at least 30% of C
Blend 2	at most 40% of A at least 35% of B at most 40% of C

Each litre of blend 1 can be sold for $1.10 and each litre of blend 2 can be sold for $1.20. Long-term contracts require at least 10,000 litres of each blend to be produced.

Formulate this blending problem as a linear programming problem and find the optimal solution using the simplex method. Validate the results by computer.

9. A large department store chain faces the problem of allocating men's suits from warehouses to various retail stores in the chain. Each store has a different price for this item. Furthermore, the shipping cost for sending suits from the

warehouse to the store depends on the particular warehouse/store combination. Additionally, there are not as many suits available as have been requested by managers of each of the stores. Information is given in Table 9 below on the selling price and costs of suits for each store, the requests of each store and the supplies of each warehouse.

Table 9. Data on selling price, costs, shipping costs, store requests and warehouse supplies of suits.

	Selling Price	Cost
Store A	$60	$35
Store B	$45	$35
Store C	$50	$35
From Warehouse	To store	Cost per Suit
1	A	$5
1	B	$6
1	C	$3
2	A	$0
2	B	$0
2	C	$4
Store Requests	Warehouse Supplies	
A – 8 suits	1 – 12 suits	
B – 16 suits	2 – 18 suits	
C – 14 suits		

10. A shipping company ships from three factories ($F1, F2, F3$) to six warehouses ($W1, W2, ..., W6$). Having a Management Science (MS) group, the company has kept shipping quantities optimal and up to date with respect to varying shipping costs. However, some months ago, the city expressed interest in acquiring the land on which one factory ($F1$) and one warehouse ($W1$) are located. Recognizing the importance of social responsibility, the company built a new factory ($F4$) and a new warehouse ($W7$) and sold the land to the city. The transportation quantities and related costs before the move are given in Table 10 below.

Table 10. Transportation quantities and related costs before the move.

	Warehouses						Factory capacity
	1	2	3	4	5	6	
Factory 1	$5	$6	$4	$5	$5	$9	8,000
Factory 2	$4	$6	$2	$4	$4	$8	4,000
Factory 3	$6	$6	$7	$5	$4	$8	7,000
Warehouse requirements	1,000	2,000	5,000	4,500	2,500	4,000	19,000

(a) Treat as a transportation problem and find the optimum shipping schedule for the present factories and warehouses using Vogel's approximation method or other suitable method.

(b) Find the optimum schedule assuming that the new factory and new warehouse have been implemented, if

   (i) the shipping costs from the new factory ($F4$) to the warehouses ($W2$, $W3$, ..., $W7$) are $9, $7, $5, $5, $9, $7, respectively; and

   (ii) the shipping costs from the two factories ($F2$ and $F3$) to the new warehouse ($W7$) are $7 and $8, respectively.

Assume that the capacity of the new factory is 9,000 and that the requirements for the new warehouse are 1,000. Again use Vogel's approximation method or other suitable method.

(c) Determine whether the total shipping cost has increased or decreased, and by how much.

SUBJECT INDEX

W

Y

Kluwer Texts in the Mathematical Sciences

KLUWER ACADEMIC PUBLISHERS – DORDRECHT / BOSTON / LONDON